PUBLIC AND PRIVATE LAND MOBILE RADIO TELEPHONES AND SYSTEMS

Lawrence Harte
Alan Shark
Robyn Shalhoub
Tom Steiner

Prentice Hall PTR
Upper Saddle River, NJ 07458
www.phptr.com

Library of Congress Cataloging-in-Publication Data

Harte, Lawrence.
    Public and private land mobile radio telephones and systems/ Lawrence Harte, Alan Shark, Robyn Shalhoub.
    p. cm.
    Includes index.
    ISBN 0-13-673609-2
    1. Trunked radio. 2. Mobile communication systems. 3. Radiotelephone. I. Shark, Alan. II. Shalhoub, Robyn. III. Title.

   TK5103.488 H37 2000
   384.5'3--dc21   99-087838

Editorial/Production Supervision: *Joan L. McNamara*
Acquisitions Editor: *Bernard Goodwin*
Marketing Manager: *Lisa Konzelmann*
Editorial Assistant: *Diane Spina*
Cover Design Director: *Jerry Votta*
Cover Designer: *Anthony Gemmellaro*
Manufacturing Manager: *Alexis R. Heydt*

Prentice Hall books are widely used by corporations and government agencies for training, marketing, and resale. The publisher offers discounts on this book when ordered in bulk quantities. For more information, contact:

Corporate Sales Department
Prentice Hall PTR
One Lake Street
Upper Saddle River, NJ 07458
Phone: 800-382-3419; Fax: 201-236-7141; email: corpsales@prenhall.com

Printed in the United States of America
10 9 8 7 6 5 4 3 2 1

ISBN: 0-13-673609-2

Prentice-Hall International (UK) Limited, *London*
Prentice-Hall of Australia Pty. Limited, *Sydney*
Prentice-Hall Canada Inc., *Toronto*
Prentice-Hall Hispanoamericana, S.A., *Mexico*
Prentice-Hall of India Private Limited, *New Delhi*
Prentice-Hall of Japan, Inc., *Tokyo*
Pearson Education Asia Pte. Ltd.
Editora Prentice-Hall do Brasil, Ltda., *Rio de Janeiro*

Acknowledgements

We give special thanks to the many gifted people who gave their expertise and personal time to help create this book. Help from these people and others was so important as land mobile radio systems use many different technologies that are often proprietary and not published. The experts below come from equipment manufacturers, service providers, trade associations, and other telecommunications related companies.

Ali Shahnami from APCO International, Inc, Dan Roszelle of Avtec, Joachim Storjohann with BESCom, Antonio Ferguson at Ce Comunica, Barbara Vonderheid from Centennial Communications Corp., Stephanie Smith, Martin Dunham and William Poellmitz of Cerulean Technology, Inc., David George with ComSpace Corporation, Douglas Bigrigg at Daniels Electronics Ltd, Greg Nohalty from Data Express, Steve Barry of EG&G Opto Electronics, W.A. Zimmerman with Electronic Communications Co., Mark Langston, Dominick Arcuri, Dave Helfrich, Joe Zwetolitz, Douglas Johnson at Ericsson, Elaine Baugh and Chris Downes from Futurecom, Pam Rentz of Glenayre Electronics, Rockie Rish with Granger, Inc., Barbara Asbury at Hewlett Packard Company, Scott Pringle and Reed Danuser from IDA, Gene Clothier of Incom Communications Corporation, Michael Joseph with International Wireless Communications, Dirk Weber, Pat Russell and Rick Bittick at Kenwood Communications, Robert Lowell from King Communications, Tom Carello of Lenbrook Wireless Communications, Roger Salmon with Macaw Electronics, Gerald Happy at Maxon America Inc., David Binns from Modular Communication Systems Inc., James Heberle of Monterrey Telecommunications Technology, Raof Khosrow, Joshua Kiem, Christine Cantarino, Christina Kalsan, Roy Bullon, Tina Castadimas, Eric Ziolko, David Kurt, Rich Barnett, and Ricardo, Bovo at Motorola, Joel Holl and Kara Palamaras from Nextel Communications, Ilkka Kupiainen of Nokia Mobile Phones, Jim Dunham with Pana Pacific Corporation, Bobby Chandler at Penta Corporation, Phil Saba from Pond & Co, Christian Weischer of Selectric, Bob VonBuelow with SetCom, Alex Cena and Karen Nielsen at Soloman Bros. Smith Barney, Andrew Trevelyan from Tait Electronics-Christchurch, Rafael Garcia of Teltronic, S.A., Bjorn Engstrom with Test-Tronics, Sean Heyen, Transcrypt, Rhett,Grotzinger and Susan McQuaid at Trident Micro Systems, Sean Johnson from TX RX Systems Inc., Cap Nguyen of Uniden, Rikki Lee and Tom Brookshire at Wireless Week, Lance Stafford from WPCS and Jim Bonnichsen with Zetron.

We also offer our sincere gratitude to the professional team of editors, illustrators and researchers at APDG Publishing, Inc.: Cindi August, Linda Oxenford, Nancy Campbell, Judith Rourke-O'Briant, James Harte and Susan Nero.

The Authors

Lawrence Harte

Lawrence Harte is the president of APDG, a telecommunications research and book publishing company. Lawrence has experience with early stage high growth companies and large multi-national telecommunications companies. His professional experience includes product management, development, marketing, design, and testing of telecommunications, radar, and microwave systems.

Between 1994 and 1995, Lawrence was Vice President of Product Management for ReadyCom where he managed development for cellular voice paging products. From 1992 to 1994, Lawrence was the New Products and Applications Manager for Ericsson General Electric. He created and managed a core team of line organization managers to develop product plans for new products. Between 1989 and 1992, he was a Digital Cellular Development Engineer at Audiovox Corporation. In this position, he was the liaison between Digital Industry Standards Committees, Toshiba's Engineering and Cellular service providers. From 1986 to 1989, Lawrence was an Automatic Test Equipment Engineer for Test System Associates on site at Westinghouse. At the Westinghouse facility, he converted technical performance requirements into electronic interface devices and created software programs for automatic testing. Between 1977 and 1985, Lawrence was an Electronics Technician in the US Navy, which included Calibration Laboratory Supervisor for microwave and RF up to 18 GHz.

Lawrence has an Executive MBA from Wake Forest University (1995) and a BSET from the University of the State of New York (1990). Mr. Harte has taught and received certificates from many non-university courses including Dale Carnegie, Tom Hopkins and Sandler Sales Institute sales, Microwave Measurement and Calibration, Cryptograph, Radar, and Nuclear Power. Lawrence has authored over 12 books and research reports on telecommunications technology, including *The Comprehensive Guide to Wireless Technologies* (1998), *Cellular and PCS, The Big Picture* (1997), *Wireless Resale Market Report and Forecast 1997-2002* (1997), *Cellular and PCS/PCN Telephones and Systems; An Overview of Technologies, Economics, and Services* (1996), *Digital Cellular: Economics and Comparative Technologies* (1993), and *Dual Mode Cellular* (1991). Lawrence is an inventor of several patents including several that are used to increase the battery life in cellular telephones (patent numbers 5,568,513 and 5,224,152).

Lawrence has consulted and been an expert witness for companies including Hughes Network Systems, Casio, Ericsson, Samsung, Sony, AMD, VLSI, Siemens and others.

Email: Lharte@APDG-Inc.com
web: www.APDG-Inc.com

Alan R. Shark

Alan R. Shark is President and CEO of the American Mobile Telecommunications Association (AMTA), headquartered in Washington, D.C. He also serves as President and CEO of AMTA's sister organization, the International Mobile Telecommunications Association (IMTA). AMTA and IMTA are the premier organizations representing the specialized wireless (SMR) telecommunications industry. Combined, the organizations have more than 400 members.

Beginning its 13th year of service, AMTA is the recognized leader for the United States SMR industry, being the largest and most influential trade association for this segment of the mobile wireless industry. AMTA is now referred to as the "Business Communications Industry Association."

Mr. Shark is a sought-after speaker and writer on issues affecting the specialized mobile wireless community. Among his various past positions, he served as Vice President of Marketing, Voice Computer Technologies Corp., and as Director of Marketing for the North American Telecommunications Association. Mr. Shark holds a Master's Degree in Public Administration and a Bachelor's in Business Administration from Bernard M. Baruch College, with further graduate work at the University of Southern California's Washington Public Affairs Center.

E-mail: ashark@imta.org
Web: www.imta.org

Robyn Shalhoub

Robyn Shalhoub is the Senior Director at the International Mobile Telecommunications Association (IMTA), headquartered in Washington, DC.

Officially launched in 1995, IMTA is the leading organization to represent and serve the commercial trunked radio (PAMR, TRS, SMR) industry worldwide. Its board of directors is internationally-balanced with representation from world-class corporations and national trade associations worldwide. Members include the largest and most well-known trunked radio operators and equipment manufacturers throughout the world.

Ms. Shalhoub joined IMTA in July 1995 as the Manager of Programs and Projects for the organization. She was promoted to Director in 1996 and to Senior Director in 1997. Her responsibilities include conducting industry research, producing IMTA publications, publishing IMTA's news report *Global Channels*, developing marketing strategies and overseeing membership recruitment. One of her major accomplishments at IMTA is the research and publication of the first comprehensive study on the commercial trunked radio industry worldwide. The report contains more than 1,000 pages of information and covers more than 50 countries. She also supervises a year-round IMTA internship program.

Ms. Shalhoub holds a Master of Arts' degree in International Communication from The American University in Washington, DC and a Bachelor of Arts' degree in International Studies from Pepperdine University in Malibu, California. Before joining IMTA in 1995, Ms. Shalhoub was project manager in the Office of General Counsel at Iridium, Inc. Prior to that she was an independent consultant to Drake Beam Morin International, Inc.

E-mail: rshalhoub@imta.org
Web Site: www.imta.org

In Memoriam
Paul Thomas Steiner, Jr. (1945-1999)

During the development of this book, tragedy happened when one of our great authors died. Tom Steiner was an expert in land mobile radio communications. He did, however, have the opportunity to add his knowledge to this book for which we will all benefit. The tribute to Tom below only provides a small glimpse of how great a man he was.

Paul Thomas "Tom" Steiner was raised in Grapeville, PA along the main line of the Pennsylvania railroad. The son of a glass worker and his wife, Tom developed a love for radio communications at a young age. He built his first HAM radio from a kit and earned his Amateur Radio operator's license from the Federal Communications Commission in 1959 at the age of 14.

Steiner graduated from Penn Technical Institute in Pittsburgh, PA, and attended classes at Point Park College and Washington & Jefferson College. He spent his entire adult life working in the Land Mobile Radio (LMR) field.

Tom worked briefly for Reliance Electric Corporation before joining the Radio Corporation of America (RCA) in 1967. His assignments included various positions in Technical Support, Engineering, and Project Management during a 17-year career.

From 1984 until 1990, he was an independent consultant, working in marketing and technical support for such firms as Cooper Power Systems, ITS Corporation, Audiovox Corporation, and Shintom Co. Ltd. During this period, Steiner was instrumental in many projects that revolutionized the LMR industry, including the vast changes in both cellular and trunking technologies.

During 1990, he became the incorporator and served as the Vice President/General Manager of Kyodo West, Inc. In 1995, he founded Professional Technology Systems, Inc. (PT Systems). These businesses marketed and sold LMR products worldwide, including developing markets in Africa and South America.

Steiner was a member of the IEEE Communications Society, a Senior Member of the Society for Technical Communications, and held a General Radiotelephone Operator License issued by the FCC in the United States. He was also listed as an Honored Professional in "Who's Who in Executives and Professionals," 1998-1999 edition.

About IMTA

Background

In 1995, IMTA was formed as a sister organization to the American Mobile Telecommunications Association (AMTA). AMTA has represented the commercial trunked radio industry in the United States for more than ten years. IMTA was created as a separate entity to effectively respond to the needs of the growing international commercial trunked radio industry and to allow for broader international participation. The organization has its own Board of Directors, membership categories and dues structure.

Vision

IMTA is uniquely organized with an international governing body designed to effectively represent the worldwide interests of the commercial trunked radio industry. IMTA's primary goal is to stimulate industry development through the creation of mutually beneficial partnerships, with entities having an active or potential international interest in the industry.

IMTA is a worldwide network of industry leaders with an interest in the growth and development of the commercial trunked radio industry. To achieve broad participation, general membership is open to operators, manufacturers and other service providers. National trade association membership is available to those associations which represent the industry in countries throughout the world. IMTA supports and fosters the creation of national trade associations in countries worldwide in order to organize local industry representation to respond to local needs.

IMTA is committed to representing the commercial trunked radio industry worldwide and to promoting the industry's growth and development. To achieve these goals, IMTA developed the following primary objectives:

- promote efficient spectrum management worldwide
- work through national trade associations to lobby for model regulations
- educate regarding spectrum use, technological advancements, and regulatory developments
- establish forums at which issues facing the industry may be addressed
- serve as an unbiased forum for discussing policies and ideas
- disseminate market research information regarding industry developments
- be a diversified organization acting as a collective force to further industry growth.

IMTA
1150 18th Street, NW, Suite 250
Washington, DC 20036
Fax:+202.331.9062
Telephone:+202.331.7773
Web:www.imta.org

"I dedicate this book to my love, Tara, and my children, Lawrence William and Danielle Elizabeth. To Tara, I love you to infinity and beyond."

Lawrence

"This book is also dedicated to the IMTA Board Members, both past and present, who shared a dream with me to create a truly international federation representing business wireless systems. Without their support and encouragement, my contributions to this excellent publication would not have been possible."

Alan

"I dedicate this work to the board of directors of the International Mobile Telecommunications Association, whose support and vision have made it possible to include the research and information on this dynamic industry."

Robyn

Foreword

Commercial Trunked Radio is currently in operation in over 61 countries worldwide, and it will continue to grow as more and more countries realize the enormous benefits to their economies.

In some countries, we refer to trunked radio as Specialized Mobile Radio, Enhanced Specialized Mobile Radio, or Public Access Mobile Radio, just to name a few. Recognizing the many different terms for essentially the same basic service, be it analog or digital, IMTA prefers to use the term Commercial Trunked Radio.

What characterizes this type of radio system apart from traditional LMR systems is not so much the technology but the nature of exclusive licenses that allow operators to operate as a carrier. Operators of private internal systems often have little incentives to upgrade equipment, maximize capacity, and develop larger coverage areas. On the contrary, commercial trunked radio operators have economic incentives to operate efficiently, expand coverage areas, and to offer an array of different services (voice & data) and more.

Today there are over 8,000,000 units in service worldwide. We use this particular measure for analysis since a typical customer may order and maintain eight radios on average. The overall growth of "units in service" is expected to increase very dramatically as we see more and more digital systems coming on line.

We are extremely pleased to have been invited to help in the preparation of this worthy endeavor, and are pleased to be looked upon as a truly international association representing the Commercial Trunked Radio Industry worldwide. This book is truly a "must read" for anyone considering entering this highly lucrative business.

Alan R. Shark
President & CEO
International Mobile Telecommunications Association
Washington, D.C.

Contents

Chapter 1

Introduction to Land Mobile Radio

The mobile wireless communications industry easily ranks as one of the most dynamic and fast-growing— if not the fastest growing—industries of today. Driving its popularity and growth are the wide variety of services it provides and the tremendous benefits it offers. Around the globe, convenience, improved efficiency, and enhanced productivity have become its trademarks.

Conventional Land Mobile Radio (Two-Way)

Conventional systems dedicate a single radio channel to a specific group of users who share it. As such, privacy is limited. It is possible for a company using a channel to be overheard by other users on the same channel. Some of these listeners might even be competitors! Conventional systems, by limiting a group of users to a specific channel, also limit the total number of customers who can be served by the system. Moreover, because radios on conventional systems transmit and receive on a single channel, the user must wait if the channel is occupied by another conversation. For these reasons, conventional systems are considered spectrally inefficient when compared to trunking systems. Figure 1.1 shows a block diagram of a conventional land mobile radio system.

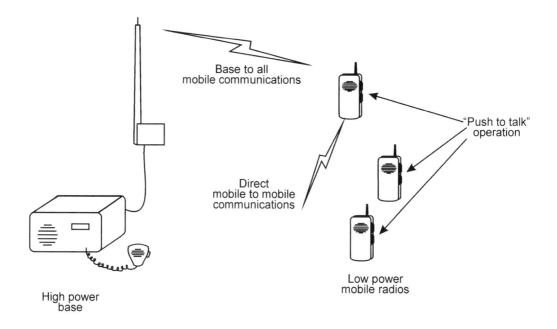

Figure 1.1, Conventional Land Mobile Radio System

Trunked Radio

Trunking systems, using frequency-trunked technology, were developed to use radio spectrum more efficiently, while offering companies a more sophisticated, private, and efficient way of communicating with their mobile workforce. Trunking systems are more expensive than conventional systems, but they also offer significant benefits and improvements in spectral-efficiency.

Unlike conventional technology, trunking allows for the automatic sharing of multiple radio channels. This means that a group of channels is assigned to a group of users who then share the channels. The advantage of this arrangement is that when a user attempts to make a call with the radio, a trunked system searches for an available channel and assigns it to the call. A different radio channel may be assigned each time the customer uses the radio; it may even switch during the same conversation. Either way, users are unaware of the swap. In the event the system is fully loaded and all channels are in use, the user either receives a busy signal or calls are "queued" until a channel is free. After the channel is selected, users have private use of the channel, which reduces interference and eavesdropping.

Trunking is considered much more spectrally efficient because switching between multiple radio channels allows less blocking and provides service to more radios per channel. Consider that on a 20-channel conventional system, roughly 700-1,000 users can be served. In contrast, those 20 channels on a trunked, dispatch-type system can service between 2,000 and 2,500 users!

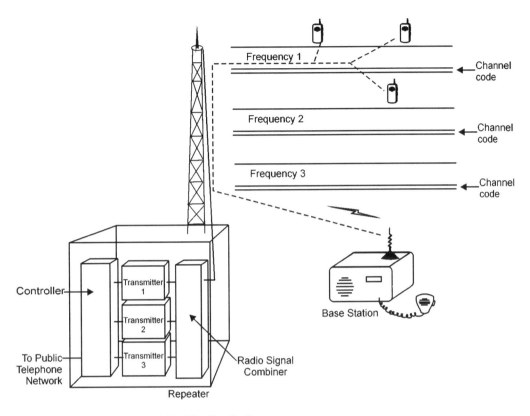

Figure 1.2, Trunked Land Mobile Radio System

Trunked systems also offer customers wider coverage areas through 1) interconnection with the public switched telephone network (PSTN), which allows trunked radio users to communicate with any user of the wireline telephone network; and 2) interconnection with other trunked systems, which may or may not be assigned to that user. Figure 1.2 shows a trunked land mobile radio system.

Commercial Trunked Radio

One relatively small, but significant, segment of the overall mobile wireless industry is commercial trunked radio, which has only recently begun to receive worldwide attention. This is because commercial trunked radio systems usually serve a very specific user group, rather than the public at large, and the major growth of the industry has occurred only within the last five years.

Today, a wide range of commercial trunked radio users exist as well as a variety of tech-nologies and services to meet their needs. As word spreads about the industry and regulators allow for it to exist, we will see commercial trunked radio systems being introduced in country after country with increasing opportunities.

The term "commercial trunked radio" was created by the International Mobile Telecommunications Association (IMTA) in an attempt to create a universal definition encom-passing the many names for the industry and to identify a specific kind of service.

As mentioned above, this small segment of the wireless communications industry has expe-rienced rapid growth primarily outside the United States within the last five years. As the indus-try is created in each country, there are an increasing number of names and classifications gov-ernments use to identify the service. For example, commercial trunked radio is known as Specialized Mobile Radio ("SMR") in the United States and is typically referred to as Trunked Radio Systems ("TRS") in Asia and Public Access Mobile Radio ("PAMR") in Europe. Figure 1.3 shows a commercial land mobile radio system.

Because the service is subject to different regulations in each country, it is difficult to create a single "name" for the service without first creating a definition. So, the following was developed.

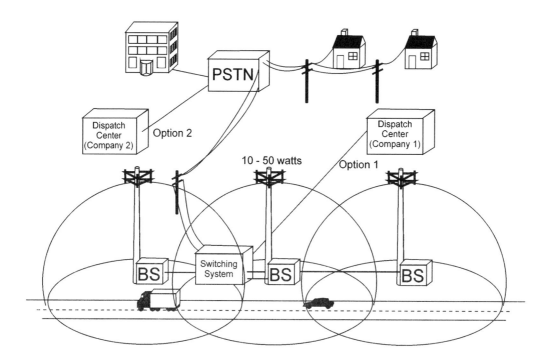

Figure 1.3, Commercial Land Mobile Radio System

The commercial trunked radio industry consists of wireless radio communications systems which employ either conventional or frequency-trunked technology to achieve two-way mobile radio communications. Most conventional and traditional trunked systems are analog systems, but several manufacturers have developed sophisticated digital technology aimed at dramatically increasing system capacity.

Commercial trunked radio systems generally provide one-to-many and many-to-one mobile wireless voice communications services, often called mobile dispatch services. Commercial trunked radio systems were developed to serve entities with a particular need for one-to-many and many-to-one mobile communications because they require the ability to communicate with large or small groups of employees on the road, especially those in fleet, sales and service operations.

Historically, commercial trunked radio technology can enable a company to simultaneously communicate with all units in a mobile and/or portable fleet or to direct transmissions to a single radio or subgroup of radios. For example, a customer can configure his trunked radio system to allow him to press a button and talk with one individual, press the same button and talk with one or more groups of people, and then press that same button again and talk with everyone on the entire network. This allows a company to operate with increased flexibility and efficiency.

Although commercial trunked radio systems are traditionally characterized by this ability to offer one-to-many and many-to-one voice communications, today they offer much more, including data capacity and full access to the PSTN. Many systems also offer integrated services such as voice mail, data messages, faxes or data transfer.

Commercial trunked radio systems are operated mostly by commercial entities, commonly called "service providers," who supply and resell their services to other entities for a profit. Commercial trunked radio systems are operated by commercial entities whose sole purpose is to provide a competitive business communications service. This is in contrast to trunked radio systems owned and operated by government agencies or by private companies for their own internal use. As more and more regulators allow entrepreneurs and businesses to operate commercial trunked radio systems, they allow a viable new marketplace to be created within their borders, with services tailored to meet the needs of particular user groups and reduced cost through competition.

The Need for Land Mobile Radio

The basic need for land mobile radio (LMR) is to allow independent two-way radio operation for private users of two-way radio services. Land mobile radio systems traditionally are used by organizations that have the need for instant communication such as taxi service, retail, construction, hospitality, conventioneers, police and fire departments. The users of these rugged communication devices, as shown in Figure 1.4, belong to a predesignated group of other users on the same account. All the radios, or a sub-group of radios can receive an instant communication by any party in the group. A push-to-talk method is used during the dispatch call or reply. This push-to-talk, radio-to-radio communication efficiently utilizes the airwaves because of the bursty nature of the information.

Locations of Commercial Trunked Radio Systems

Since 1990, changes in regulation and increased demand have led the commercial trunked radio industry to be introduced in a number of new markets throughout the world. At the end of 1997, it is estimated that there were commercial trunked radio systems in at least 55 countries, particularly in Europe, Latin America and the Asia-Pacific region, serving an estimated 6.8 million users. The rapid growth trend is predicted to last at least until the year 2015.

Figure 1.4, Traditional Land Mobile Radio
Source: Motorola

North America

In North America, the first country to introduce the commercial trunked radio industry in 1974 was the United States. It was at this time that the U.S. Federal Communications Commission (FCC), the primary telecommunications authority, recognized the need for greater and more efficient mobile communications. The FCC formulated two new types of mobile services, cellular and Specialized Mobile Radio (SMR), the U.S. term for commercial trunked radio. The FCC created separate rules and regulations for the two industries, expecting each to  attract a unique group of users. SMRs were classified as "private carriers" and, as such, were not subject to "common carrier" regulations at either the state or federal levels. Cellular services, on the other hand, were classified as "common carriers" due to their expected appeal to the public.

The FCC also provided, and continues to provide, discreet spectrum for private, internal-use trunked systems.

The first U.S. commercial trunked radio network was launched in 1977. A decade later found approximately 3,000 operators with 628,000 units in service. By the end of 1997, the number of subscribers exceeded 2.68 million. The largest operator in the world was found in the United States: Nextel Communications. This company had more than 2.2 million subscribers at the beginning of 1999.

The constant growth of the U.S. industry is attributable to the growing importance of wireless communications, favorable economic conditions, the affordable price of most commercial trunked radio services and, most importantly, a positive regulatory environment. Continued growth is expected as enhanced services are introduced and more spectrally efficient technologies are implemented.

In Mexico, the government has been taking steps to deregulate the telecommunications sector to allow for private investment and to improve the quantity and quality of communications services. The deregulation process began in the late 1980s and, as new regulations were implemented, the market for commercial trunked radio, cellular, and paging was opened to private investment. By the end of 1996, 45 companies were operating commercial trunked systems in Mexico with an estimated 73,000 units in service.

The commercial trunked radio industry has been active in Canada since 1984. It is estimated that there are more than 50 and perhaps even hundreds of companies operating such systems today. The Canadian Wireless Telecommunications Association estimates that the total number of subscribers on these systems is 200,000. Some consolidation has occurred in the industry recently, which has left it with about three large operators and many small operators throughout Canada.

Europe/Eurasia

Until the late 1980s, most two-way radio systems in Europe were private networks owned and operated by utility companies, taxi companies, public safety groups, private companies, and government entities. At that time, deregulation in Europe, coupled with the desire among many telecommunications companies to offer new services, led to the introduction of the commercial trunked radio industry in the region. The first system was introduced in 1987 in the United Kingdom. In 1990, the French government issued its first commercial trunked radio licenses, and systems following soon after, known as "Chekker networks," appeared in Germany. Many other countries soon followed suit.

For those European and Eurasian countries that gained independence in the late 1980s/early 1990s, an inadequate public telephone infrastructure and the desire of the new regimes to privatize the communications sector led to liberalization and the immediate introduction of commercial trunked radio services in many markets.

Today, more than 175 companies are operating commercial trunked radio systems in Belarus, Belgium, the Czech Republic, Estonia, Finland, France, Germany, Latvia, Lithuania, Moldova, Poland, Portugal, Romania, Russia, Spain, Sweden, Tajikistan, Turkey, Ukraine and the United Kingdom. The largest markets are found in France, Germany and the United Kingdom.

In some of these countries, such as the United Kingdom and Germany, there may be hundreds of additional smaller systems in operation—just how many is unknown. In other European countries—including Bulgaria, Denmark, Hungary, Italy, Luxembourg, the Netherlands, Norway, and Switzerland—government agencies still control the trunked radio systems in operation. However, in many of these countries, regulators are investigating ways to introduce commercial trunked radio services in their markets over the next few years.

Although there were fewer than 100,000 commercial trunked radio subscribers in Europe in 1992, that number topped 567,000 in 1997 and is expected to increase to 1 million by 2000.

Central and South America and the Caribbean

The commercial trunked radio industry has seen increased growth in Latin America particularly in the last five years. This growth can be attributed to the region's high demand for reliable two-way voice communications and for business tools that improve productivity and efficiency. This demand has been able to be met due to the decision among many governments to privatize certain telecommunications services. This has opened a variety of telecommunications services to competition, including commercial trunked radio. Today, about 250 companies have licenses to operate these systems in Argentina, Brazil, Chile, Colombia, Dominican Republic, Ecuador, El Salvador, Guatemala, Honduras, Nicaragua, Panama, Peru, Trinidad & Tobago, and Venezuela.

Given that the commercial trunked radio industry has only been active in Latin America for a short time, the number of licenses that have been issued is significant. However, it should be noted that not all of these licensees are operating systems. In fact, in 1995 it was estimated that only 200,000 units were in service in these countries. With such high demand for service, the Latin American market is severely untapped. As more systems are fully implemented and others emerge over the next decade, we can expect to see increased industry growth.

The largest markets in the regions are Argentina, Brazil, Chile, Colombia and Venezuela, each with more than 10,000 total subscribers. The smaller markets have anywhere from a few hundred to a few thousand subscribers. However, competition is heating up in each market. Evidence of consolidation has appeared in Argentina, Brazil and Colombia, where operators are banding together to capture a larger market share. Many operators also are implementing larger systems to obtain larger capacities to serve the growing demand for these services.

Asia-Pacific

The commercial trunked radio industry has existed in Asia-Pacific since 1982, when the first system went into operation in New Zealand. Soon thereafter systems were also introduced in Japan and Australia. In most other countries in the region, however, the service was either relatively unknown or offered only by government agencies. In the last five years, many Asian-Pacific countries have experienced high economic growth, which has prompted demand for sophisticated business communications services. As governments have liberalized the telecommunications sector, commercial trunked radio has been able to meet this demand. Today, more than 250 companies have licenses to operate these systems in Australia, Cambodia,

China, Hong Kong, India, Indonesia, Japan, Korea, Malaysia, New Zealand, the Philippines, Singapore, Sri Lanka, Taiwan and Thailand.

Considering the industry's recent debut in the region, the number of licensees is significant. It seems safe to say that the industry has yet to experience its major growth spurt. In fact, licenses were issued for the first time in 1996 in major markets, such as India and South Korea. Operators anticipate a huge demand for services in these countries as well as several other Asian markets at least over the next five years. However, one factor that may limit commercial trunked radio growth in the region is the prohibition of operators to provide interconnect service. This restriction is found in several Asian countries. It is hoped that this restriction will eventually be lifted or relaxed as operators and associations such as IMTA continue to pressure governments to adopt flexible regulatory policies.

Africa/Middle East

The commercial trunked radio industry has also appeared in several countries in Africa and the Middle East in the last few years, although growth in these two regions has not been as extensive.

While many countries in Africa are in need of reliable two-way communications systems, many governments are currently focusing on industry development and unification rather than creating a thriving telecommunications sector. While some radio systems are in operation, they are commonly operated by government entities or United Nations forces rather than private companies. The key exceptions are the Ivory Coast, South Africa and Uganda.

In the Middle East, demand for voice communications as well as business communications tools are in demand. Highly centralized governments have been hesitant to release radio frequencies to the private sector, although some are now considering doing so. The exceptions are in Turkey, where nine companies operate commercial trunked radio systems; Lebanon, where 67 companies operate these systems; and Jordan, where the government recently issued the first tender for commercial trunked radio systems.

As countries in these regions stabilize and open their telecommunications markets, the commercial trunked radio industry will have many opportunities for expansion and growth.

Commercial Trunked Radio Services

Commercial trunked radio systems were created to provide mobile dispatch voice services via analog systems. As demand for commercial trunked radio services has increased, so has the demand for more enhanced and diversified services. To meet this demand, technologies have been developed that expand and improve analog services. For example, trunking technology improved analog service by allowing more privacy, channel capacity and spectrum efficiency.

While service rules vary from country to country, additional enhanced and diversified services have evolved such as interconnection to the PSTN, networking between sites, messaging, and the ability to transmit and store various forms of data.

Conventional systems ordinarily provide:
· on-site direct radio to radio communications;
· wide-area coverage in remote areas;
· more efficient utilization of frequencies;
· enhanced dispatch capabilities;
· increased radio coverage.
As briefly explained above, trunked systems offer additional capabilities, including:
· improved system access;
· call privacy;
· user priority levels for system access;
· more efficient frequency utilization;
· flexibility in assigning multiple talk group levels;
· enhanced dispatch capabilities;
· system management capability.

The differences between analog and digital technology are substantial. For instance, operators who use digital technology claim that this technology is able to provide end users with the following:
· increased communications reliability;
· increased capacity to handle more calls;
· greater call privacy;
· reduced exposure to fraud;
· easier multiplexing;
· easier encryption;
· better monitorability of performance;
· integration of switching and transmission;
· opportunities for data and voice services, such as paging, messaging, and voice mail.

Voice

The most well known application for land mobile radio is wireless voice communications. Voice communication can be basic telephony or two-way voice (radio-to-radio). The service rates charged for land mobile radio voice services are normally lower than service usage charges for cellular systems. For private systems where the radio license permits private radio operation to be owned by a company, there are usually no service fees aside from the equipment maintenance costs. When LMR operators provide service for public use (such as for a taxi cab company), a flat monthly fee is charged and there is no usage fee. When the land mobile radio system offers telephony service, a fee is commonly charged for connection to the public telephone network since an operator must pay for the wired telephone line (and sometimes the usage of the line). When LMR systems provide voice services that interconnect with a public telephone network, they ordinarily charge per minute usage fees for both two-way voice and basic telephone service.

Dispatch

Dispatch radio service normally involves the coordination of a fleet of users via a dispatcher. All mobile units and the dispatcher can usually hear all the conversations between users in a dispatch group. Dispatch operation involves push-to-talk operation. Common dispatch service rates are shown below.

The commonality between dispatch service and mobile telephony is that communication occurs for brief periods. There is usually no charge for the airtime usage for SMR systems, but ESMR systems commonly charge usage fees for dispatch service.

Paging and Messaging

Paging and messaging services allow users to send and receive text messages up to approximately 200 characters per message. Messages can be cascaded to allow the sending of longer messages.

Depending on the type of system, the charge for unassisted messaging services varies from "no cost" when bundled with other services (such as voice) up to 50 cents per message. The cost per message ordinarily comes from operator assisted messaging.

Data

One of the fastest growing areas for land mobile radio is data service. Data services transfer information between computers and data terminals. Examples of data devices used in land mobile radio systems include computers used in police cars and digital dispatch terminals used in taxi cabs and delivery trucks.

A new type of operator console and service has been created to take advantage of data services on LMR systems. This service is called computer aided dispatch (CAD). CAD systems are a computerized communication system that can coordinate and/or track mobile vehicles. CAD systems can comprise various degrees of complexity, from automated messaging devices to complex computer systems that display maps and vehicle positions on a computer monitor.

Similar to the messaging fee structure for LMR service, the fees charged for data services range from "no charge" to over 10 cents per kilobyte.

Commercial Trunked Radio Standards

Although several standards/protocols have been developed, there is not one universal standard for commercial trunked radio systems. Within the industry, there are a mixture of key protocol standards that have been developed by manufacturers in the United States and Europe.

Analog Systems

In the United States, where the FCC has not promoted or created a standard, there are three analog protocols, each of which was developed by a manufacturer in the mid-1970s: E.F. Johnson Co., Ericsson Inc., and Motorola Inc. These manufacturers and many others, such as Kenwood and Uniden, supply equipment to commercial trunked radio operators throughout the world.

In the case of the United States, while market forces have dictated industry development and created a diverse and robust industry, the absence of a standard has led to the production of equipment that is incompatible with dissimilar systems. Over the last ten years, the emergence of multiple vendors for the Johnson LTR® protocol has been critical. The adoption of this protocol by so many vendors has led to what some consider a "de-facto standard." Furthermore, groups of operators in the U.S. are now creating wide-area networks through consolidation, purchases of trunked systems or roaming agreements with other operators.

In Europe, the MPT 1327 protocol for analog trunked systems was pioneered by several manufacturers and first published in 1985. The first MPT systems went into operation in 1989, and MPT standard equipment is widely used throughout Europe and in many other countries. Many variations of the standard have been adopted. One example is Autonet, a Finnish standard based on MPT 1327. Many European operators and policy makers believe that, while creating a standard has contained market development, it has also helped the industry grow by creating flexibility while retaining competition.

Other companies, such as SmartLink Development L.P. and ComSpace Corp., are producing enhanced new technologies for commercial trunked radio systems. SmartLink's SMRLink® technology is geared toward wide area networking and Unique's DCMA technology is aimed at increasing system capacity without obtaining additional channels.

Digital Systems

To meet the growing need for advanced services and to lower system equipment costs, the land mobile radio industry is migrating to digital radio technology. Two digital mobile radio standards are being developed by association organizations in the U.S. and Europe. Other proprietary digital protocols have been developed by private companies. The first, Terrestrial Trunked Radio (TETRA) is being developed by the European Telecommunications Standards Institute (ETSI), and the second, APCO-25, is under development by the U.S. Association of Public Safety Communications Officers (APCO).

TETRA is seen as a replacement for the analog trunked MPT 1327 systems and large conventional analog systems. It is structured similar to MPT 1327 trunked analog products, in that interfaces are standard. However, within the infrastructure, each vendor will implement different interfaces to meet the different needs of end users. Forty-six organizations worldwide have signed a Memorandum of Understanding to support TETRA. To date, several TETRA systems are already in operation throughout Europe and several operators, such as Motorola and Nokia, have reported additional system orders.

APCO-25 is targeted specifically for the public safety and security sectors. Because these sectors usually are served by government operated or private systems, it probably will not be used for commercial trunked radio services.

While only two standards for digital radio technology are being developed, several companies have developed digital technology for this new and growing market. Some examples are EDACS® by Ericsson Inc. and iDEN® by Motorola Inc. Ericsson reports that more than 350 EDACS® systems are in operation worldwide, including Brazil, Canada, Hungary, Italy and the United States. Motorola reports that eight iDEN® systems are in operation worldwide in, Argentina, Canada, Colombia, Israel, Japan, Singapore, and the United States. In addition, the company currently is installing additional systems in Argentina, Brazil, China, Korea, Mexico, and the Philippines. The new generations of land mobile radio handsets have similar features to cellular telephones along with advanced traditional LMR features (e.g., dispatch).

Trunked Mobile Radio Equipment

Commercial trunked radio units are either hand-held portable radios or mobile radios installed in a vehicle. Many newer units look exactly like cellular phones and others look more like telephone handsets. Some units look like traditional land mobile radios. However, end-user equipment is changing rapidly. New generations of both analog and digital trunked radio units offer built-in paging, messaging, phone directories, voice mail, data, and more. These units offer advanced ergonomic designs created for more consumer appeal.

Commercial Trunked Radio Users

The most common customers of commercial trunked radio systems are those with a need for one-to-many and many-to-one radio communications with mobile workers. With the introduction of more sophisticated technology and the development of enhanced and diversified services, a broad spectrum of end-users now exists. Following is a partial list:

· Electrical Suppliers and Distributors;
· Gas and Water Distributors;
· Bus Companies;
· Airport Authorities;
· Railways;
· Courier Services;
· Emergency Services;
· Construction Companies;
· Trucking and Fleet Management Companies;
· Agriculture Industry;
· Health Services;
· Dispatch Companies;
· Taxi and Limousine Companies;
· Manufacturing Industry;
· Petroleum Industry;
· Service Technicians;
· Insurance Agents and Adjusters;
· Government Agencies;
· Real Estate;
· Security Services;
· Plumbing & Electrical Contractors;
· Concrete and Asphalt Companies.

Commercial Trunked Radio Equipment Suppliers

There are far fewer equipment suppliers worldwide than there are system operators. Designing and manufacturing equipment requires a lot of time, money and a skilled workforce. In many countries, the licensing mechanism is put into place before local companies have the time and the financial resources to design their own systems, so operators entering the market tend to purchase equipment from established suppliers. Examples of major equipment suppliers are Alcatel, Celwave, E.F. Johnson, Ericsson, IDA Corp., Kenwood, King Communications, Motorola, Nokia, Simoco, Securicor, SmartLink, Standard, Tait, Uniden, Yaesu and Zetron.

Many additional companies manufacture terminals, rather than system equipment. In addition to the companies listed above, examples of major terminal equipment suppliers are: AEP, Aiko, Alinco, Andrew, Aselsan, Bendix, Cleartone, DeTeWe, Icom, Kavicom, Key, Maxon, Rohde & Schwarz, Sinclair, Talco and Teltronics.

Trunked Radio Competition

Because commercial trunked radio was designed to offer one-to-many and many-to-one communications to the business community, it represents a unique and specifically targeted industry segment without direct competition. There are, however, three mobile wireless technologies that provide similar services and may seem to be competitors: cellular telephony, paging and personal communication services (PCS).

Cellular telephony is not yet a direct competitor to commercial trunked radio because it does not provide the same kind of communication and is designed to meet the needs of a different end-user group. Cellular telephony, in contrast to commercial trunked radio, was developed as an outgrowth of the PSTN to serve the public by providing one-to-one communications. Using this system, a mobile unit can call another mobile unit or a wired telephone through interconnection with the PSTN and vice versa. Cellular systems are beginning to offer dispatch services (e.g., the group call feature of the GSM system). However, most cellular system infrastructure equipment is unfriendly to one-to-many dispatching needs. Many countries also limit the ability of commercial trunked radio systems to compete with local telephone service by prohibiting them from interconnecting with the local telephone network.

As commercial trunked radio develops worldwide and appropriate regulations are implemented, there is some overlap with cellular service. For example, when interconnection is allowed, commercial trunked radio systems allow for one-to-one communication with users of the wired telephone network. For this reason, some consider it similar to cellular. But commercial trunked radio offers its customers much more. Personal communications service (PCS) is similar to cellular service in that it is toward two-way personal communications.

Paging is another form of wireless communications that is not directly competitive to commercial trunked radio. Paging serves individuals and businesses and allows data messages to be sent instantly to a mobile unit. Because the message traditionally delivered to the pager is a one-way numeric or alphanumeric message, the person carrying the pager can respond only by using a telephone. As a result, paging is less convenient than instantaneous two-way voice mobile communications and offers neither one-to-many nor many-to-one services but regularly acts as an adjunct to either cellular or commercial trunked radio. The newer land mobile radio systems do offer the ability to send short messages directly to the handset. This allows land mobile radio systems to better compete with paging systems.

Radio Spectrum Regulation

Because radio spectrum is a finite resource, rules and regulations must be in place to manage its allocation and to ensure that it is being used efficiently. At the same time, these regulations often determine the availability and types of services that can be offered in any given market and can decide whether commercial trunked radio will flourish or stagnate in a given country.

Spectrum regulation occurs on an international level and on a national/local level. At the international level, the regulatory body charged with allocating spectrum is the International Telecommunications Union (ITU), created as part of the United Nations. The ITU constitution states that it will

a) affect allocation of bands of the radio frequency spectrum, the allotment of radio frequencies and the registration of radio-frequency assignments in order to avoid harmful interference between radio stations of different countries;
b) coordinate efforts to eliminate harmful interference between radio stations of different countries and to improve the use made of the radio-frequency spectrum [ITU Constitution, Article 1, Section 2].

In carrying out these tasks, the ITU allocates frequencies to the broadest extent possible so that national governments have the flexibility to make more specific allocations that are appropriate for their markets. For example, the ITU allocates frequencies for "land mobile communications" rather than "trunked radio" so that each national or local regulatory body can allocate specific frequencies to industries and services that fall under the category of "land mobile communications." While specific allocations rest with the national/local governments, allocating widely accepted frequencies is beneficial because it promotes the growth of similar services in various countries.

Membership in the ITU is voluntary, as is participation in its meetings and complying with its recommendations. However, the ITU does serve a very significant purpose in that it is a forum at which national telecommunications authorities and other interested parties can convene to discuss agreements and/or disagreements regarding telecommunications issues.

Specific frequency allocation and the creation of rules and regulations associated with the licensing and operation of systems are left to the telecommunications authorities in individual countries, which may be national or local in scope. Regulatory bodies may be government-run ministries of telecommunications, semi-independent organizations, totally independent, or self-funded organizations. In many countries, the regulator is also still the operator of telecommunications services. However, many countries have separated these functions.

Frequency spectrum allocations for commercial trunked radio vary from country to country. Figure 1.5 shows the division of frequency use in the United States in 1998 for private radio users.

Developing rules and regulations is critical and necessary to ensure spectrum efficiency and the protection of users. Toward this end, national or local telecommunications authorities have numerous decisions to make regarding license allocation methods, application procedures, fees, license requirements, etc. Each decision influences the way in which the industry will mature. For example, if a government does not establish appropriate regulations requiring licensees to build and operate their systems, companies can receive a license but never build the system or provide service to the public. This can potentially keep sincere operators out of the market and limit the industry's ability to grow and develop. Moreover, it prevents the business community from using the beneficial services that commercial trunked radio provides. In addition, some countries require licensees to pay such high licensing and annual fees that they often do not have enough money to implement the system and provide the service. When entering a new market it is essential to consider the range of regulations for trunked radio service in the target country. Everything from loading and construction rules, equipment standard requirements and interconnection regulations, may have a significant impact on a commercial trunked radio business.

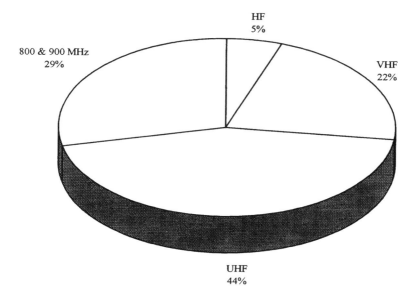

Figure 1.5, Division of Frequency use in the United States in 1998 for Private Mobile Radio Source: Strategis Group, Washington D.C.

Summary

The commercial trunked radio industry has achieved tremendous growth within the last five years. As countries throughout the world continue to open their markets, numerous opportunities for further expansion abound. The diversity of markets, regulatory environments and customer needs throughout the world create an interesting and challenging landscape for the industry. Indeed, the prevailing trend toward liberalization and competition should prove to give the industry an ideal environment in which to flourish.

Chapter 2
SMR Technology and Evolution

Commercial trunked radio is different from Cellular and Personal Communications Services (PCS) in that it has its roots in "push-to-talk" radio systems. Push-to-talk can also be referred to as half-duplex (or simplex), whereas the half means only half of a conversation can happen at one time. This is sometimes experienced when using a household speakerphone. Full duplex, on the other hand, allows two simultaneous conversations, similar to a classic argument between a coach and a sports official. Push-to-talk communications range from a simple pair of hand-held citizen band (CB) walkie-talkies to a new breed of enhanced, full-duplex, cellular-like consumer products. Specialized mobile radio (SMR), a term for this service used by the Federal Communication Commission (FCC), is a two-way radio system that is operated by a private company who resells use of that system to other companies for a profit. SMR is primarily used by the transportation and service industries. Though classified by the FCC as a commercial mobile radio service (CMRS), some SMR systems are not currently classified like cellular and PCS as a common carrier telephone system. Private mobile radio (PMR) is a private two-way radio system used for internal communications only. PMR is typically used by the public-safety, emergency services, utilities and government agencies sectors.

It is possible to connect SMR systems to the public switched telephone network (PSTN). This is accomplished by attaching a special interface adapter to the base station. During this process, the SMR radio channel is connected to a two-way regular telephone transmission line. The wired portion of the call extends the effective distance of the radio communication link. Even though a two-way wired connection is made, the communication itself remains a half-duplex, however, some subscriber models are full-duplex allowing a "cellular-like" use.

Mobile Radio Systems

There are several types of mobile radio systems. These include conventional radio systems, trunked specialized mobile radio (SMR) and enhanced specialized mobile radio (ESMR).

Conventional Two-Way Radio

Conventional two-way mobile radio systems allow communication on a dedicated radio channel between two or more mobile radios. Traditionally, conventional two-way radio systems have served public safety and industrial applications. These systems, as a rule, only allow push-to-talk service (one-way at a time) to allow the sharing of a single radio channel frequency.

Until the early 1990s, many conventional two-way radio system communication protocols were unique to the manufacturer of the radio equipment. Because the protocols (radio language) used in conventional two-way radio systems have become more standardized, the selection of the type of two-way radio system is now more determined by cost, reliability, coverage distance, and type of services desired.

Conventional two-way radio systems are available in a wide range of frequency bands. Natural properties of radio waves indicate, as a rule, that lower frequencies provide longer ranges. Antenna selection and usage also effect the optimization of these conventional systems.

A very popular commercial two-way service called dispatch allows a base station operator to broadcast over the entire coverage area. Often, selective signaling squelch tones or digital signals are added so that only a select group of radios will hear each other. For example, ABC Taxi Service uses a dispatch service, but selective signaling allows only ABC Taxi cabs to hear the voice announcement. In some cases, all group members hear the responding mobile, and in other cases only one group member makes a private response back to the base. These basic systems are mostly analog half-duplex or simplex technology. The name two-way to describe this technology is misleading because conversations travel only one-way at a time. The single channel in question, however, sees traffic going two ways, base-to-mobile and mobile-to-base.

Because conventional two-way radio systems allow many users to share a large geographic region, it is typical that interference can occur between users that are operating close to each other. Receiving all conversations on the radio channel would be annoying. For this reason, squelch systems were developed. Squelch is a process where the audio output of a radio receiver is controlled (enabled) by the reception of an incoming RF signal that is above a predetermined level. Squelch allows a radio user to avoid listening to noise or interference signals of distant radio transmissions that occur on the same frequency. Squelch systems are almost always based on carrier level, and tone or digital code signaling can be added to improve this feature.

Carrier controlled squelch systems mute the audio until the incoming radio signal is above a specific signal level. Carrier controlled squelch allows a radio user to avoid listening to noise or weak interference signals of distant radio transmissions that occur on the same frequency. Unfortunately, any strong incoming radio signal will pass through the squelch system. This includes nearby users whose conversations may not be of interest to the user. To overcome this limitation, tone controlled squelch systems were developed.

Tone controlled squelch systems mute the speaker of a radio receiver unless the incoming radio signal contains a specific tone. Tone controlled squelch allows a radio user to avoid listening to noise or interference signals of all other radio transmissions that occur on the same frequency and that do not have the correct tone mixed in with the audio signal.

Digital squelch systems are very similar to tone controlled squelch systems. Digital squelch systems mute the speaker of a radio receiver unless the incoming radio signal contains a specific digital code. Digital controlled squelch allows a radio user to avoid listening to noise or interference signals of all other radio transmissions that occur on the same frequency and that do not have the correct code mixed in with the audio signal.

Figure 2.1 shows a typical conventional two-way radio system. In this diagram, several mobile units that are operating within a radio coverage area can share a single high power radio channel carrier. To help differentiate between different groups of users, a separate squelch tone signal is mixed in with audio signals for specific groups of users e.g., a taxi company and a flower delivery service. This squelch tone is ordinarily a frequency that is below the audio frequencies so the users cannot hear the tone.

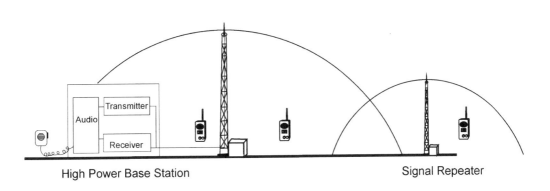

High Power Base Station   Signal Repeater

Figure 2.1, Two-Way LMR System

Trunked Radio

Trunked radio is a system that allows mobile radios to access more than one of the available radio channels in that system. If a radio channel is busy, the mobile radio can access any other channel in the system that is not busy. This type of system allows mobile radios to find open channels during busy periods.

Many of the newer Specialized Mobile Radio (SMR) operators have converted their conventional radio systems into a group of radio channels (called a trunk group). Mobile radios that operate on trunked radio systems will automatically choose any radio channel in the trunk group that is unused at the moment. This technique greatly decreases the amount of blocked calls compared to the use of dedicated independent radio channels. Most trunked SMR equipment still operates only in the half-duplex mode, which requires push-to-talk operation by the SMR user.

Trunked systems have several available radio channels. Mobile radios that operate in this system and that wish to communicate may search for an available radio channel by looking for identification tones or digital control messages. Optionally, some trunked radio systems use dedicated control channels to coordinate access to available radio channels.

Private mobile radio (PMR) systems traditionally allow communication between a base and several mobile radios. PMR systems can share a single frequency or use dual frequencies. Where PMR systems use a single frequency, mobile radios must wait to talk: this is called a simplex system. To increase range and improve system efficiency, some PMR systems use two frequencies: one for transmitting and another for receiving. To simplify the radio design, some models cannot transmit and receive at the same time: this type of unit is called half duplex. When PMR systems use two frequencies and the subscriber radios can transmit and receive at the same time, they are called full duplex.

SMR systems commonly provide advanced services not possible with traditional PMR systems. SMR systems always use radio channel pairs, whether or not the subscriber units operate half-duplex (requiring push-to-talk) or full-duplex. Although direct mobile-to-mobile communications is possible (called "talk around"), it is not the usual mode of operation. Talk around is useful if the subscriber units leave the coverage area of the system or in case of system failure. Originally, SMR systems were limited to a single tower site. Linking between systems was difficult, with roaming between systems by a given subscriber being very difficult. Telephone interconnect has always been technologically feasible but not originally allowed by regulatory agencies and not always used by SMR operators. More recent technological innovations have provided for advanced features such as multi-site networks, roaming capability, etc. SMR systems typically cover a range of about 25 miles or more by using a single tower, high building or mountain top antenna site.

Figure 2.2 shows a typical SMR system. In this system, one or more dispatchers communicate to mobile radios in a relatively large geographic area (normally a city). A single SMR service provider may provide service to truck delivery companies, repair technicians, courier companies, and other groups of users, so each group of users must have their own identification codes. Mobile radios communicate by coordinating with the nearest SMR tower and transmitting requests for service.

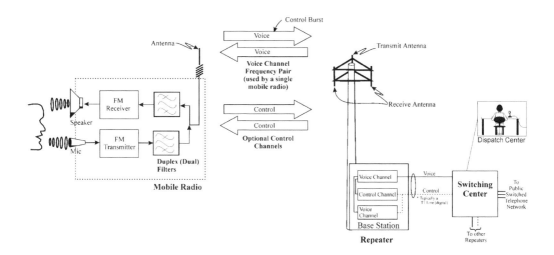

Figure 2.2, SMR System

Enhanced Specialized Mobile Radio (ESMR)

Enhanced Specialized Mobile Radio (ESMR) appears almost exactly like a cellular phone system. ESMR is basically defined by who is providing the service, more so than the actual technology being employed. Older SMR systems were owned by individual entrepreneurs serving limited areas. Interoperability between systems, say between Motorola, Ericsson/GE and EF Johnson (now Transcrypt) systems, was not possible. ESMR is regularly associated with a new class of SMRs that use digital radio technology to provide advanced telecommunications services to customers. These services include telephone, dispatch, numeric and alphanumeric paging, fax and data services. A good example of the new breed of ESMR is the Nextel system. From a customer's point of view, Nextel ESMR service is similar to traditional cellular service because it also provides one-to-one communications services.

A key benefit of the ESMR systems is the combined push-to-talk feature along with the full duplex telephone operation packaged into one handheld portable telephone. The user of this combined handset has the choice of using the push-to-talk, unit-to-unit feature or using it as a "cellular" phone that is capable of full duplex interconnect service. Priorities can be given to public safety (e.g., State Police, Fire and Rescue) to prevent consumer ESMR users from loading up the system during public emergencies.

Like cellular, most ESMR systems have a complex of cells allowing frequency reuse and increased system capacity. ESMR calls are all routed to a Mobile Switching Center (MSC) just like that in a cellular system.

Most enhancements for the ESMR system are accomplished by digital radio transmission. There are several digital radio technologies employed for ESMR systems, including TDMA and frequency hopping spread spectrum (FHSS). The leading digital ESMR systems include integrated Digital Enhanced Network (iDEN®) developed by Motorola, Enhanced Digital Access Communications System (EDACS©) developed by Ericsson and Terrestrial Trunked Radio systems, a protocol developed by the European Technical Standards Institute (ETSI) and manufactured by several companies. These digital systems are designed to service large numbers of radio users and provide telephone, dispatch, and data services that compete with traditional cellular and PCS services. There are also other emerging technologies such as DC/MA, which is being developed by ComSpace Corp. and which promises to operate from a single site and achieve an 8 to 1 capacity gain over analog.

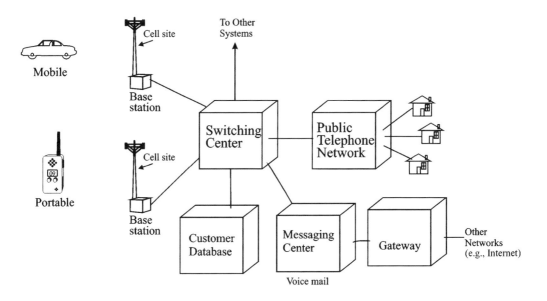

Figure 2.3, ESMR System

In most ESMR systems, there are many mobile radios. Most ESMR systems have many cell sites (base stations) in a geographic region that can simultaneously transmit and receive on several radio channels at the same time. Each cell site has a radio coverage radius of about 5 to 15 miles. The systems usually provide service to large geographic regions of several cities or even nationwide service. Many are interconnected.

Figure 2.3 shows a sample ESMR system. In this system, ESMR mobile radios communicate with the system by first requesting access to the network on a control channel. The system responds to a valid request with a radio channel assignment. The ESMR system also has a switching system that can connect ESMR mobiles to each other or to other networks such as the public switched telephone network. The mobile switching center in an ESMR system must operate very quickly to allow two-way, push-to-talk operation. When an ESMR customer presses the "push-to-talk" button, they must be connected to other radios in their group within about ¼ second or less. This is very different than a cellular phone system where the setup of a telephone call can take 20 to 30 seconds.

Technologies

Land mobile radio systems are converting from traditional analog systems to digital systems that offer integrated voice and data services. The key types of communication channel structures used in LMR include simplex, semi-duplex, and full-duplex. Repeaters are used to extend the range of LMR systems. The technologies used by LMR systems continue to evolve to allow for advanced services and more efficient systems.

Analog

Many of the traditional two-way radio systems use analog radio technology. Most of these systems use FM modulation; however, there are some systems that use single sideband amplitude modulation. Single sideband modulation reduces the amount of frequency bandwidth required for each radio communications channel. Until recently, single sideband AM modulation suffered from excessive signal distortion. New modulation and signal processing techniques have overcome many of these limitations, and equipment is now available that uses single sideband technology.

Figure 2.4 shows a typical analog two-way radio system. In this diagram, a mobile radio has a FM transmitter and FM receiver bundled together. Because the mobile radio is used for "push-to-talk" service, the antenna does not need to be connected to both the transmitter and receiver at the same time. When the user presses the push-to-talk button, the antenna is connected to the transmitter. When the button is released, the antenna is connected to the receiver. A squelch circuit is connected to the receiver to allow the receiver audio to reach the speaker only when the receiver level or tone code is of sufficient level to ensure a good received signal. When the level is below that threshold (normally set by the user via a squelch knob), the speaker is disconnected and no sound (such as noise) can be heard. The mobile radio communicates with a radio tower that has a high power base station. The base station is usually connected to a control console that allows a dispatcher to communicate remotely from another location (such as an office) by using a telephone line connection.

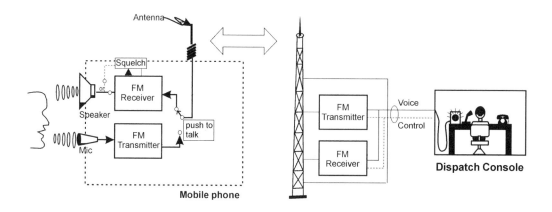

Figure 2.4, Analog Two-Way Radio Technology

Digital

Many of the newer two-way mobile radio systems use digital radio technology. Most of these systems use some form of digital modulation. The use of digital transmission allows for many new advanced services.

Figure 2.5 shows a typical digital two-way radio system. In this example, a digital mobile is connected to a data display in a mobile vehicle. The radio transceiver has a transmitter and receiver bundled together. Some digital systems combine the control channel with a traffic channel while others use dedicated control channels. Digital systems can either use a single digital radio channel for each user or can divide a single radio channel into time slots for use. The use of time division allows for several users to simultaneously share each radio channel.

Most digital land mobile radio systems can provide for short message services (SMS) and data transfer. These services allow numeric and alphanumeric paging and data transfer. Enhanced digital systems allow for handoff from one radio channel to another. This allows for cellular-like operation. To allow for advanced services (such as messaging and radio channel handoff), control messages and data may be sent between the base station and the digital mobile. Control messages may command the digital mobile to adjust its power level, change frequencies, or request a special service (such as three way calling). To send control messages while the digital mobile is transferring digital voice, the voice information is either replaced by a short burst (blank and burst) message in some systems (also called fast signaling), or else control messages can be sent along with the digitized voice signal (called slow signaling).

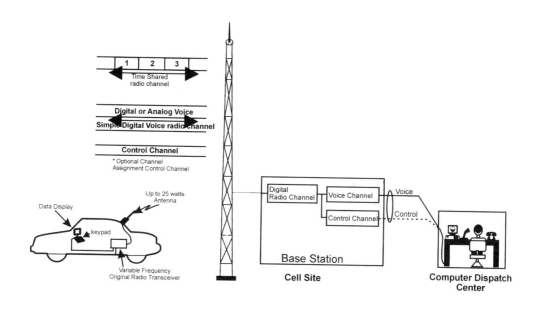

Figure 2.5, Digital Land Mobile Radio Technology

Simplex Systems

Figure 2.6 shows the simplest of conventional stations. Simplex systems consist of a receiver, transmitter, control system, antenna switch, and antenna system. The receiver and transmitter operate on the same frequency. The operator is equipped with a microphone, loudspeaker and push-to-talk switch. Each operator is a controller (coordinator) of the radio system.

While the mobile radio is in the receive mode, the antenna system is connected to the receiver by the antenna switch and conversation on the radio channel can be heard by the operator. When the operator desires to speak, the push-to-talk switch is pressed. This action causes the receiver to be switched off, the transmitter to be switched on, and the antenna system to be connected to the transmitter by the antenna switch. With this type of operation, only one side of the conversation can be heard at a time. With one transmitter in operation, all other receivers in the system are operating. To reply, a given operator must wait until the push-to-talk switch at the transmitting station is released.

For on-site communications, only the types of radios described above are required. Communications are carried out directly between individual stations. This somewhat limits communication range however, especially between hand-held portable radios, which typically have low power transmitters in order to conserve battery life.

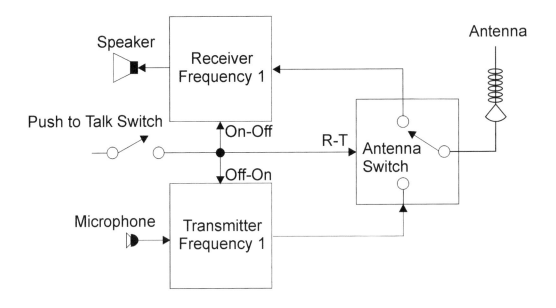

Figure 2.6, Simplex Station Block Diagram

When increased area coverage is required, one of the simplex stations is equipped with a higher power transmitter and an improved antenna system. This station (commonly called a "base station") serves as the "dispatch point" for the mobile and portable radios in the field. In many instances, the base station equipment has a tall antenna tower and is located on a hill or mountaintop. Because radio transmission at higher frequencies (e.g., UHF) is often characterized as "line of sight," this enhances the communications range of radios that operate in the system.

If a dispatch office is in a different location from the antenna tower, the radio equipment can be configured to provide control system functions via "remote control." This remote control scheme can use either wired connections (typically telephone lines) or radio links.

Semi-Duplex Systems

Another type of a simple communication system is the "semi-duplex" system. The semi-duplex system operates in a similar method as the simplex system, except that the receiver and transmitter operate on different frequencies. In the semi-duplex system shown in Figure 2.7, the mobile only communicates with the dispatch radio on a single frequency and it must listen to the dispatcher on a different frequency. It is not possible to transmit and listen at the same time. In this type of system, the dispatcher has a transceiver that can simultaneously transmit and receive. This type of operation is used in very busy dispatch environments, such as taxi companies. Chatter between individual mobile units would unnecessarily tie up valuable "air time" that the dispatcher could use to coordinate customer pick up schedules.

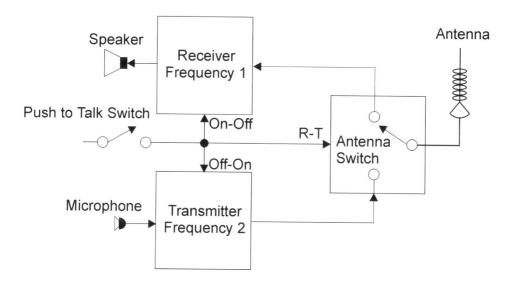

Figure 2.7, Semi-Duplex Conventional Station

Full Duplex Systems

With a full duplex station, both the receiver and transmitter can operate simultaneously on different frequencies. This means that both sides of the conversation can be heard at the same time by both users. Full duplex systems are commonly used for radio telephone (mobile phone) type services.

Figure 2.8 shows a full duplex station. In this system, the antenna is connected to both the transmitter and receiver through a frequency "duplexer." The frequency duplexer contains two bandpass filters. Each filter only allows a band of frequencies to pass through. One filter allows the receiver frequencies to pass through and the other filter allows the transmitter frequencies to pass through. The filters keep the high power transmitter signal from entering into the receiver section.

Full duplex conventional systems are often used in rural telephone systems. These systems operate on a "party line basis," using a single radio channel pair. Another use is for emergency call boxes, such as on Interstate highways and even computer controlled light rail vehicles. In these cases, the user of the remote full duplex station simply picks up a handset and automatically receives either a "dial tone" or is directly connected to the dispatch station.

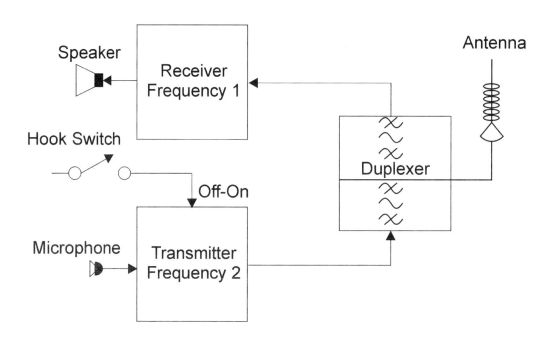

Figure 2.8, Full Duplex System

Repeater Systems

A popular method of extending the range of radio system communications is through the use of "repeaters" (also known as a "mobile relay"). Repeater stations (Figure 2.9) re-transmit radio signals on a different frequency on a relatively high power transmitter.

Typically, the repeater station is located on a hill, mountaintop or tall building with an efficient antenna system. The mobile and portable units in the system operate in the "semi-duplex" mode. One drawback to this mode of operation is that the mobiles and portables cannot communicate directly with each other. Should the repeater be inoperative due to various reasons (e.g., a lightning strike) or should the mobiles or portables leave the communication range of the repeater, they would become inoperative. For this reason, they are often equipped with the ability to switch to the "simplex" mode and communicate directly with each other. This feature is also known as "talk-around."

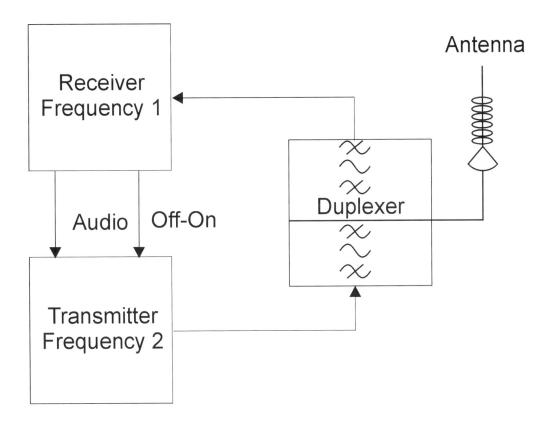

Figure 2.9, Repeater Station Block Diagram

Figure 2.10, Mountain Top Repeater System
Source: Daniels Electronics

Repeaters are often mounted in rural areas. These rural areas can have extreme environmental conditions. Figure 2.10 shows a repeater system that is mounted on a mountaintop.

Community Repeater Systems

A specialized type of repeater system is known as the "community repeater." In a community repeater system, multiple talk groups are possible on the same radio channel pair. This is accomplished by some sort of selective calling capability. In the United States, among other countries, a Continuous Tone Coded Squelch System (CTCSS) is often employed. Each talk group is assigned a specific squelch code. If the correct code is not received from the repeater, the receiver remains muted (quiet). Under normal operating conditions, a given user can only hear conversations between other members of the same talk group.

Community Repeaters are the forerunner of early trunked systems in that they permitted multiple talk groups to share infrastructure resources. Many community repeater systems remain in operation today, mainly in suburban and rural areas, servicing small companies with a mobile work force.

Conventional System Enhancements

Many enhancements have been added to conventional radio systems. These include telephone interconnect, simulcast radio transmission, automatic unit identification, selective calling, scanning receivers, data transmission and even GPS receivers with mapping capabilities, among many others. Wide area systems have been employed, and simulcast techniques have been used to enhance radio coverage.

Conventional systems do not adapt well to multiple channel operation. Even though mobile and portable stations can be equipped with scanning receivers and multiple channels, channel selection is left to the user. In high density, multiple channel environments, this can result in missed calls, long delays in connect time and ineffective spectrum management.

Key Applications

Public Safety

Public safety users include local government, police, and firemen. These users ordinarily require communications in a relatively defined geographic area (such as a city) between defined groups of users. Recently, public safety users have begun to use advanced devices to integrate information sources. These services include fingerprint recognition, instant access to police databases for criminal records, scanning of licenses and other forms of identification for database recognition, etc.

Emergency

Emergency users include medical technicians, ambulance and rescue workers. These mobile workers require voice and data information to be transferred reliably. Services commonly used by emergency service organizations include typical two-way radio voice communications to coordinate rescue efforts and data messaging to transmit key medical information. Vehicle tracking devices can be used by emergency service personnel to coordinate the location of emergency vehicles.

Industrial

Industrial users include workers involved in the manufacture or production of products or services. These include the petroleum industry, the agricultural sector, steel production and other light and heavy industry workers. These sectors normally use two-way radio services for their instant communications capability and for the remote control and monitoring of equipment.

Two-way radio systems often provide plant personnel with monitoring devices to analyze certain plant conditions. For example, a petroleum company will often implement two-way radio monitoring devices along its pipelines to monitor levels of petroleum and leaks. Farm workers may use two-way radio systems to monitor irrigation and flood warning levels, fence penetration, livestock tracking, etc. Figure 2.11 shows a product that is capable of being connected to existing monitoring equipment. The EL2000 produced by EchoPort, Inc., allows the connection of a monitoring device to a radio that has the ability to wirelessly transmit information or alerts directly to a system or an end-user (e.g., plant manager) when a pre-specified parameter has been exceeded. This device can be attached to sensors for temperature, light, humidity, motion, power and status monitoring. Popular areas for deployment include server rooms, home security systems and industrial or manufacturing environments.

Utilities

Power plant, water, and gas utility companies often maintain their own private communication systems. These utility companies may have large geographic areas with many groups of workers. With the consolidation of utility companies, there is a need for compatible systems in large geographic regions. Similar to the industrial users, the communication systems must be reliable.

Figure 2.11, Remote Monitoring Device, EL2000
Source: EchoPort, Inc.

Recently, utility companies have begun to integrate information services into the communication systems. These include access to documentation (such as maps and service manuals) and remote sensing and control of external devices such as electric heaters and shutoff valves. Electric companies are also installing remote sensor devices so that they can detect household power usage for billing purposes. Gas companies use remote monitoring devices to monitor gas levels in pipelines. Water companies use these devices to monitor water levels and ensure that equipment is working properly.

Transportation

Transportation industries such as railroad, trucking, and highway maintenance have found many uses for two-way radio systems. Vehicle tracking is particularly important in the transportation sector in order to monitor the location of merchandise which the vehicle is carrying. By monitoring the vehicle's location, risk of theft and high-jacking can be minimized. Transportation companies also use two-way radio systems for their voice communications capabilities in order to stay in constant contact with the home base (in case a delivery is called off or the drop off location is changed) and with other mobile units in case assistance is needed. Highway maintenance personnel use two-way radio services to determine traffic levels on roads and to coordinate construction projects. Railroads use two-way radio not only for essential voice communications, but also to monitor track performance and interference.

Marine

Marine users range from government agencies, such as the U.S. Coast Guard, to private ships and port authorities. Each, of course, uses two-way radio systems for essential voice communications. Ships and port authorities must communicate to coordinate port entry and exit and to avoid accidents. The U.S. Coast Guard uses government frequencies to monitor ship locations and to respond to emergencies. Port authorities and other organizations also use radio devices to monitor weather patterns through terminals placed on water beacons.

Local Dispatch

Taxis, courier services and repair personnel use two-way radio systems to receive and coordinate their assignments. In the past 10 years, the trend has been for the automation of dispatch services. Taxis can run credit card charges through data terminals located in their cabs. Courier companies can provide more efficient service because their couriers are in constant contact with dispatchers. Couriers no longer have to arrive at a pick up only to learn that the order was canceled. The message can be relayed instantaneously to a two-way radio, which saves time and money. While couriers used to operate two-way radio systems primarily for voice services, now radio carriers are also using radio systems for data services. Dispatchers can send data messages to couriers regarding pick up locations, contact names, and levels of urgency. Repair personnel have found two-way radio services invaluable. Service calls can be transmitted either by voice or data so that repair personnel do not have to use the customer's telephone to find out where the next appointment is. In addition, the one-to-many and one-to-one communications capabilities are invaluable when repair technicians need advice on a particular problem or need to find a repair technician with a needed spare part.

Limousine services and hotel vans are also using two-way radio services to their advantage. Hilton Hotels has a service where a person can check into the hotel by using a data terminal located in the hotel limousine or van.

Digital dispatch display is used for a taxi service. As pickup requests are received, they are forwarded to taxi cabs close to the area. Taxi drivers can bid on which fares they want to pick-up. These messages are automatically coordinated reducing the burden (and cost) of the central dispatcher.

Business Radio Services

Business radio services are used to coordinate workers within a small company area. These uses include event management (trade shows and concerts), building maintenance personnel and mobile workers. There are numerous examples, including the Olympics, where trunked radio services are used to coordinate events. In these cases, instant communications are invaluable as workers coordinate activities and respond to emergency or urgent situations. Building mainte-nance personnel, ranging from building engineers to hotel staff, find two-way radio service ben-eficial in coordinating activities in order to provide better service to their clients.

Location Tracking

Vehicle location tracking is used in the transportation sector to track the location of vehi-cles and equipment in order to avoid accidents, to monitor the movement of merchandise and to provide more efficient delivery service. Vehicle location tracking provides for better fleet man-agement and security of staff and products that are transported. Because the actual position of a vehicle can be monitored in a central station, emergency messages can be sent from the mobile radio, together with the actual position of the vehicle. This allows rescue support to be dis-patched immediately.

Vehicle location tracking is usually performed through the use of global satellite position-ing (GPS). GPS systems receive radio signals from several satellites that are at known positions. The distance from the satellites is determined by comparison of the radio signals to determine their relative timing difference. This precise timing is used to calculate the location of the vehi-cle relative to the position of the satellites. By knowing the exact position of the satellites at a particular moment (provided by the radio signal), the position of the vehicle on the ground can be determined. This position information can be reported back to a central dispatch system through an LMR radio via a data message transmission.

The accuracy of GPS receivers is typically accurate 10 to 50 meters. By using a previous-ly known reference signal from other transmitter sites from other GPS receivers within a radius of 200 to 300 km, the accuracy of a GPS location can be improved to typically 1 to 5 meters. This is called differential GPS (DGPS).

Figure 2.12 shows the Plettac Grundig MT-118 integrated GPS receiver and land mobile radio. This device is used to determine the exact location of a vehicle and provide this informa-tion back to a central control terminal. This information is used to monitor valuable goods trans-porting, emergency call reporting, public transportation tracking and public safety monitoring.

Figure 2.12, Position Location LMR Radio MT-118
Source: Plettac Grundig

Other position location tracking (terrestrial systems—land based) can also be used. Shipping has used land based radio signal positioning systems (e.g., LORAN C) for many years. Any of these systems can be integrated with LMR transceivers to provide position location information back to a central dispatcher.

Environmental

Companies and government agencies involved with environmental monitoring and control include forestry, wildlife preservation, earthquake monitoring, water testing, pollution monitoring, surveying and others. The common thread shared amongst most of these environmental agencies is that they require monitoring communications in areas that do not have wired communications.

Security

Wireless devices can be used to enhance the security of areas that are not commonly located near wired telephone systems. These devices can include motion sensors and video cameras. Figure 2.13 shows a wireless motion sensor. When the motion sensor is triggered, an alert notification is sent over a wireless network to a central control center. That alert is immediately re-directed according to pre-specified instructions. The message can be sent to a phone, a pager, an e-mail address or a web site.

Digital cameras can be connected to wireless devices to transmit high quality still images directly to the Internet via the wireless network in real-time or non-real time (allows for data compression). These images can be used for monitoring of construction sites, rest areas along highways, automatic teller machines (ATM) and other areas that can benefit from security monitoring. Figure 2.14 shows a digital camera that is connected via a wireless system.

Figure 2.13, Wireless Motion Sensor, EM 1000
Source: EchoPort, Inc.

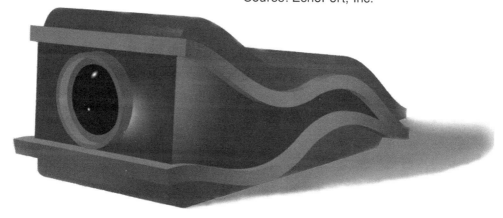

Figure 2.14, Wireless Digital Camera
Source: EchoPort, Inc.

Chapter 3

Analog Land Mobile Radio Systems

Land mobile radio (LMR) system technology is primarily divided into conventional and trunked radio systems. Conventional systems require users to share a single channel. Trunked systems allow users to automatically access a radio channel from a group of radio channels.

Analog LMR systems use Frequency Modulation (FM) or Amplitude Modulation (AM) to communicate voice information. Most commercial analog radio systems use a control channel that transfers digital information.

There are many types of analog LMR systems. Many of these are proprietary designs that may be produced by one or a limited number of manufacturers. Despite the proprietary nature of analog LMR systems, some analog systems have become standard (or popular) in the marketplace. These systems include ClearChannel LTR, SmartSite, SmartNet, SmarTrunk, SmartZone, OmniLink, LTR-Net, Passport, ESAS, APCO 16 and MPT1327. Some of these systems are only designed for single site, and others may be expandable for wide area (multi-site) conventional or trunked systems.

Logic Trunked Radio (LTR)

Logic Trunked Radio (LTR) was developed by the E.F. Johnson Company. LTR uses sub-audible signaling and distributed control logic to provide for fast system access and voice channel assignments to any pre-assigned radio channel. LTR has become the de facto standard for analog trunked radio systems. The first trunked radio system was Clearchannel LTR®.

The Clearchannel LTR® system allows the automatic sharing of channels in a multiple repeater system. The LTR system uses distributed control, where the mobile radio shares the intelligence for radio channel access and assignment with the repeater (base station). Access on the LTR system can be made on any pre-assigned RF channel that is idle.

A transmitting repeater RF channel provides information to mobile radios by simultaneously transmitting information in a low bit rate data stream (sub-audio at 150 Hz, 300 bps) that coexists with voice information. This eliminates the need for a separate dedicated control channel. Each repeater maintains its own data stream and handles all accesses on its channel. In the LTR system, radio network access control is independently handled by the mobile radios. Repeaters that are transmitting constantly provide updated information about the status of the repeater (its ID), calls that are to be received by mobiles and the status of other repeaters at the site (e.g., if they are idle). Figure 3.1 shows a basic LTR system block diagram.

Every mobile radio used in an LTR system must be programmed for LTR signaling in that system. This information includes which RF frequencies to use and the home system identification number. Preprogramming the LTR radio allows the operation of LTR mobile radios to be much simpler than that of conventional radios. This is because many functions that would normally be performed by the user are automatically performed by the LTR system. The user only has to select which system to use (or group ID) and press the push-to-talk switch. If the system is not busy serving other mobile radios, speaking can begin. If the system is busy, an audio signal (beep tone) is usually provided.

Mobile radios only transmit and receive ID codes that have been programmed into the radio by the system operator. The use of unique ID codes for each radio or groups of radios disables other mobile radios from eavesdropping on conversations (e.g., competing taxi cab company). When the incorrect ID code is received, the audio for the mobile radio is muted. Although audio could be monitored by a non-LTR transceiver (such as a scanner), frequency transmission can occur (and usually occurs) on different RF channels. It would be difficult to eavesdrop on a complete conversation.

When in dispatch mode, LTR radio systems immediately release channels after audio communication is completed (release of the push-to-talk button). This eliminates wasted hang time that could occur in dispatch (mobile-to-mobile) calls when there is no voice activity. When the LTR system is used for mobile telephone operation, the radio channel is held for the entire length of the telephone call.

Conventional systems that have dedicated frequencies experience almost no delay time when communications begins. Because LTR mobile radios must find an available radio channel prior to transmitting, this adds delay time. The LTR system keeps the delay time to a minimum by programming the mobile radio to search for an idle RF channel from a limited predefined list of frequencies and by providing a list of idle channels via the data stream of a transmitting repeater. This results in a typical delay time of less than 100 msec for LTR systems as compared to delay times exceeding several seconds for other types of trunked systems that require the use of a control channel.

Because each radio channel can be independently accessed and controlled, LTR systems can use all the radio channels for voice communications.

Basic LTR systems do not allow access priorities to be assigned. Access priorities help to control access to a busy system by only authorizing service to users with a high priority code (e.g., police or rescue). For LTR systems, mobile radios do not attempt to access the system until a channel is available. The mobile that then acquires the channel is the one that makes the first access attempt. This is a first-come-first-served method of access. For the LTR system, all mobiles have equal access priority.

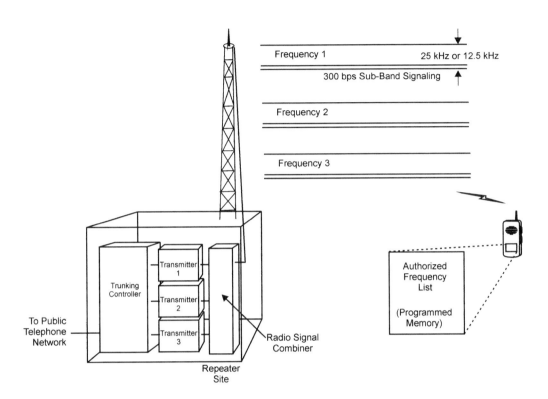

Figure 3.1, Basic LTR System

LTR mobile radios have simple controls that include power, volume control, system select, and, optionally, a group select switch. No squelch control is required because the squelch is internally controlled.

LTR mobile radios are preprogrammed with a code for their "home" repeater. The mobile receives most of its control information from the home repeater. When a mobile radio is idle (not placing or receiving a call), it continuously monitors its home repeater site to determine if it is being paged (called) and what other repeater channels are idle.

The mobile radio initially attempts to use the home repeater to make or receive calls unless it is busy. When the home repeater is busy, any other available repeater in the site may be used.

The LTR system allows up to 250 ID codes to be assigned for each repeater. The combination of each ID code and home repeater number are the "address" of mobiles in the system. Therefore, up to 1000 separate mobile radios can be addressed in a four repeater system. ID codes are assigned for individual or groups of mobile radios, as necessary.

When a mobile initially transmits, a data "handshake" is performed with the repeater. After the mobile radio transmits a service request, the repeater responds with a message back to the mobile indicating that the mobile has successfully accessed the system. The total amount of time necessary for the handshake is less than 30 msec. The handshake process prevents a mobile from capturing a channel that may have been already in use. If the handshake response is not received, the mobile radio must inhibit transmitting and attempt to access the system again.

During a transmission, data messages are continuously transmitted between the repeater and the mobile. These messages can be heard by all the other mobiles that are monitoring that channel (e.g., part of the group call). The information that is contained in the data messages is dependent on whether it is transmitted by the repeater or by the mobile radio. Each message takes approximately 130 msec to transfer at 300 bits per second.

Each data message contains fields that synchronize, identify, and transfer control messages and data. These fields include Sync, Area, In Use or Go-To-Repeater, Home Repeater, (mobile or group) ID Code, Free Channel and Error Check Bits.

The Sync field contains the first two bits of a data message. This indicates to the receiver a message is to be received and provides an indication of the timing of the message (synchronization).

The Area field is a one bit field that identifies if the message has been received from a nearby channel that may be interfering (similar to a color code in a cellular system). If the area field of the transmitted message does not agree with the anticipated area code, the message is ignored.

The In Use or Go-To-Repeater field contains the identification number of the repeater for which the message is to be transmitted. This message changes to the turn-off code message (31) when the transmitter is about to stop transmitting (mobile user has released the push-to-talk button). The use of the 31 code directs all the other mobiles in the group that are monitoring the channels (group call) to mute their audio. This prevents a noise burst (called a "squelch tail") that would occur after a transmitting mobile unkeys.

The Home Repeater field indicates the number of the assigned home repeater number.

The ID Code field contains the mobile ID or group ID number of the mobile(s) that are being called.

If other mobiles assigned to that repeater have been trunked to other repeaters to receive a call, additional messages are transmitted containing the ID codes of these mobiles. The repeater gets this information from the repeater data bus. The Go-To information described earlier tells these mobiles which repeater to switch to.

The Free Channel field indicates the number of a repeater channel that is not busy at that site or nearby sites and is available for service. When this slot is transmitted by the mobile radio, this slot contains (31) which is a pass code. This slot contains (0) if all repeaters are busy.

The Error Check Bits field is used to check for errors that may have been received in a data message. If the error check bits indicate an error has occurred, the message is ignored.

Up to 20 LTR radios can be installed at a repeater site. Each repeater monitors the status of all the other repeaters at the site. If a repeater is busy, the home repeater and ID code of the mobile that is receiving the call is continually placed in the appropriate time slot for transmission so other mobile radios can monitor the status. This allows other radios that have the same group ID code to be able to monitor the audio (late entry). If no repeater number has been assigned, the time slot has nothing inserted.

When a repeater detects that its number is being used by another repeater at the site (e.g., a home repeater was busy and another repeater was used), it begins transmitting an additional data message to indicate to its home mobiles to go to that other repeater to receive a call. This message continues to be transmitted for the duration of the call.

Data messages that are transmitted on a home repeater are time sequenced. Every third message contains information pertaining to the mobile radio that is currently receiving a call on that repeater. Slots between these messages contain information relative to mobiles that have been trunked to other repeaters.

LTR allows a validator security feature to be used to check the repeater ID code combinations. If the home repeater and ID code combination is determined to be not valid (not authorized for service), the repeater can transmit a turn off code (31) which disables the invalid mobile.

When a mobile is idle (in standby), it continually listens to data messages from its home repeater. The mobile checks to see if it is being called (paged) by another mobile and to determine which repeater is available to place a call.

SmartSite™

SmartSite™ is a Motorola single-site system that can be expanded up to five channels. This system can service approximately 500 users.

SmartWorks™

SmartWorks™ is a Motorola system that can provide up to six voice communications channels plus a control channel. The SmartWorks™ central controller can handle hundreds of users, provide for up to 2000 talkgroups and support the APCO 16 industry standard.

Motorola uses various ways to control channel assignment in an LTR trunked system. The basic Motorola system has five channels, with one channel dedicated to control signaling. The data channel transmits at 3600 baud. Each mobile initially scans for the control channel when it is first turned on. The control channel data stream contains the system ID. If the system ID assigned to the mobile radio (stored in its memory) matches the system ID that it has received on the control channel (home repeater), the mobile will stay tuned to this channel and continuously receive messages. It is listening for incoming calls (pages) and updated (changed) system information.

If a conversation is in progress for a group of users that are associated, the group code is continuously transmitted on the radio channel. This allows the mobile to decode the group ID and join in (un-mute the receiver) on the conversation in progress even after the group call has been set up.

When a mobile desires to transmit (e.g., the push-to-talk button has been depressed), the mobile sends an inbound signaling word (ISW) message on the data channel to request service. An ISW contains the Fleet number, Sub Fleet number, and mobile Unit's ID code. If the system is able to receive and process the request, it sends an outbound signaling word (OSW) that instructs the mobile to tune to an available (unused) repeater channel. If all channels are busy, the user will hear a busy tone.

After a mobile has successfully tuned to an assigned channel, it sends a 3600 baud handshake. The system then communicates with the mobile on the voice channel by sending a 80 msec tone burst at 1800 Hz. This tells the mobile that it is on the voice channel. The data rate of the handshake is then changed from 3600 baud to 150 baud sub-audio signaling (below the audio band). The 150 baud signal contains the Fleet and Sub Fleet information. The system will then send all other mobiles that are part of the Fleet and Sub Fleet groups to that voice channel. When these mobiles tune to the new channel, they decode the fleet and Sub Fleet data message and unmute their speakers so they can receive the voice message. The transmitting mobile now starts to send a sub-audible connect tone at (105 Hz) simultaneously with the voice signal. While the system hears this connect tone, it will hold the channel (continue to transmit) for that sub-fleet until the mobile unkeys. When the mobile does unkey, it sends a 200 msec disconnect tone at 164 Hz.

The disconnect tone also starts a channel hold timer in the system. Because one or more of mobiles assigned to the group are likely to respond to the initial voice conversation, the system holds the channel available for a quick response. Each time any mobile that is part of the group that has tuned to the voice channel keys up, it only needs to send a connect tone on the voice channel to respond. This tone tells the system that one of the mobiles in the group desires to respond. The system allows the transmitter to operate and resets the hold channel time-out timer. When all the conversations on this channel are done from this group, the channel hold time-out timer will eventually expire. This will completely release the voice channel.

A fleet is the entire group of mobile units and portables. Subfleets are smaller groups that are part of a fleet. For example, a subfleet of users can be maintenance people and a base dispatcher. Another subfleet could be a delivery department and their base dispatcher. The number of subfleets accessible by a single mobile can be restricted by the system operator. The number of fleets is limited to 20 channels with 305 fleets.

SmartNet™

SmartNet™ is a Motorola system that is designed to serve a large number of users from a single site with enhanced reliability and security features. Each SmartNet™ trunked radio system can have up to 28 channels and can serve thousands of users with up to 4000 talk groups. Both analog and digital voice is possible with the SmartNet system. SmartNet™ allows for simulcast transmission that uses multiple sites operating on the same frequency to extend the coverage area.

SmarTrunk™

SmarTrunk™ is one of the worlds most popular VHF/UHF trunking protocols. This analog system can operate as a dispatch, radiotelephone, or conventional system. Up to 16 trunked channels can be used per system with 4,000 subscriber capacity for roaming applications.

When operating as a radiotelephone, it uses pre-origination (store and send dialing) similar to cellular telephones. Priority assignments are possible with overrides. It also allows for remote "Radio Kill" to disable illegal or non-paying customers.

SmarTrunk II™ was first introduced in 1994. SmarTrunk II™ trunking protocol provides automatic channel assignment and call privacy. The system can support up to 16 channels to service approximately 4,000 mobiles. The system supports interconnect, group call, emergency calls, and conventional operation.

The intelligence of the SmarTrunk II™ system is a digital trunking controller. The controller performs all the control and administration, including mobile identity validation and billing record storage.

SmartZone™

SmartZone™ is another LMR system developed by Motorola. This system allows up to 48 sites, with up to 28 channels per site, to be interconnected to provide wide area coverage. SmartZone™ allows integration of conventional and trunked systems.

SmartZone™ OmniLink™ is a network option that allows multiple SmartZone™ systems to be interconnected into one very wide area communications system. The OmniLink™ option provides for up to 192 sites to be interconnected.

LTR-Net™

LTR-Net™ is an evolution of LTR technology that allows networking of LTR systems and that was developed by E.F. Johnson company. Early LTR systems were limited to systems that had one radio site. LTR-Net™ offers wide-area dispatching, automatic roaming, unique IDs with increased security, multiple frequency (cross-band) operation and over-the-air system management.

The LTR-Net™ system interconnects networks of multiple sites. Each site is capable of independently operating. However, each site is also linked to all other sites through the LTR-Net™ network switching system.

The LTR-Net™ system can be operated from a Windows NT™ computer. This allows for advanced services (such as information services) to be developed on a standard personal computer (PC) platform.

The LTR-Net™ system is backward compatible with existing LTR systems. This allows for the gradual migration of users from existing LTR systems to advanced LTR-Net™ equipment and advanced features. LTR-Net™ radios can even communicate on the same repeater as an LTR system.

LTR-Net™ system components include mobile radios (with new software), repeaters (some repeaters can have a Universal Repeater Interface), and network switches (new for LTR-Net™). The switch is the communications management center for the repeater sites. The switch transfers calls between repeaters and connects to the public telephone network when necessary. Figure 3.2 shows an LTR-Net™ system block diagram.

The LTR-Net™ switch connects repeaters to each other and other networks (e.g., PSTN or PBX). A Subscriber and System Manager (SSM) controls the switch and all the other networked components in the LTR-Net™ system. The SSM can be a personal computer that controls the switch. The SSM is the central manager for the system that includes assigning talkgroups, disabling radios, and interfacing to billing systems. The LTR-Net™ system can have one or more than one SSM.

Because LTR-Net™ is an enhancement to the existing LTR protocol, LTR-Net™ systems can co-exist with other LTR equipment.

The LTR-Net™ system allows up to 65,000 user identification codes (UIDs) for assignment to each mobile in a system. The unique mobile identification codes allow for direct radio-to-radio calling and interconnect radiotelephone services. UIDs also allow autonomous registration that permits internetwork roaming. When LTR-Net™ mobiles move into a visited system, they automatically identify themselves to the new LTR-Net™ system. The visited site registers its new mobile with the series by checking with the home system. The home system updates its database with the new location of the mobile.

LTR-Net™ provides over-the-air system programming and management tools. The LTR-Net™ system provides for the automatic disabling mobiles (e.g., non-paying customers) through a "sleep" or "kill" message that can be sent through the radio channel. When a "sleep" message is received, the radio disables the user interface and then waits for a wake-up message. When a "kill" message is received, the unit will completely disable itself and servicing will be required to reactive the mobile. Programming the authorized radio channel list can also be performed over the air.

LTR-Net™ enhances the original LTR architecture to add more group IDs. LTR-Net™ provides up to 4800 group IDs per site, independent of the number of installed repeaters. Global GIDs can also be assigned that work throughout the entire system, allowing mobiles to access any site.

Each LTR-Net™ mobile contains an electronic serial number (ESN) in addition to a unique identification number (UID). The UID is used to provide autonomous registration that allows the system to track an individual radio's location within the LTR-Net™ communication system.

For complete details about LTR-Net™, call E.F. Johnson at 1-800-228-0226 in the US and 305-591-2130 internationally.

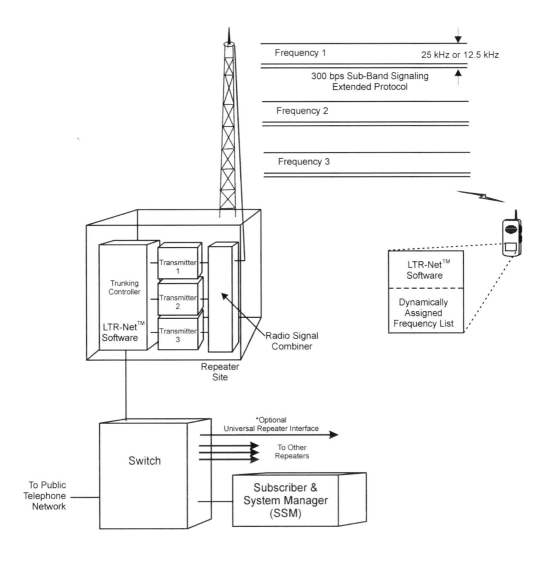

Figure 3.2, LTR-Net™ System

Multi-Net® II

A wide-area trunking system was developed by E.F. Johnson for large public safety and private systems. The field-proven components of Multi-Net® II systems around the world provides a solid foundation for LTR-Net™'s networking technology.

PassPort

PassPort is an all new industry standard trunking protocol developed by Trident Micro Systems for SMR (Specialized Mobile Radio). The PassPort system allows for advanced features that allow SMR operators to compete against other integrated wireless networks (e.g., cellular, PCS and digital LMR). The PassPort protocol allows for Wide Area Dispatch and Interconnect networking with seamless roaming between networked sites. It includes automatic registration and de-registration for mobile radios. Transmissions occur only at network sites where mobile radios are registered in that radio network group. Automatic registration allows for the most efficient use of radio channel airtime.

The heart of the PassPort system is the Network Trunk System (NTS). The NTS system is a distributed logic digital network infrastructure. The NTS provides extraordinary flexibility to system planners by eliminating the need for a centralized switch and the expense of "back-hauling" signals from each site in the network to a common switch location. In the NTS environment, each site employs an independent switch that need only be connected to the nearest site in the network. The NTS is easily scalable and may be expanded to meet even the largest operators growth requirements. The PassPort trunking protcol builds on the LTR protocol to address the need of radio users with wide-area dispatch networking, seamless roaming, selective call capability, positive system security, short message service, voice mail capability, radio stun, radio kill and other features.

Each mobile radio that operates in a passport system is assigned a unique identification number called "Mobile Identity Number" (MIN). In addition, to enable dispatch services, mobile radios that are members of a talk group are also assigned one or more Group IDs (GRP-ID). The combination of MIN and GRP-IDs provide for over 60,000 individual user ID Codes per system.

In addition to a MIN and GRP-ID identifier, each subscriber unit manufacturer imbeds a unique electronic serial number (ESN) into each mobile radio. This ESN cannot be changed by the dealer or radio user. The mobile radio's ESN can be read on-command from the system to validate the identity of a mobile radio at any time.

Because PassPort is an enhanced low speed protocol building on LTR, NTS networks supporting PassPort are LTR compatible. This allows for gradual migration of systems to advanced features and services.

Upgrading LTR systems to PassPort primarily involves replacing the existing repeater controllers. Many types of repeaters can be connected to the NTS using the standard repeater connections.

Because PassPort protocol uses an enhanced LTR type signaling to support wide area group dispatch, it allows for short call setup times (typically less than 400 msec). PassPort also can provide for privacy between two mobiles involved in a Selective (direct one-to-one) Call.

PassPort adds improved levels of security over standard LTR mobiles by performing over-the-air validation using the radio ESN and assigned MIN. When the ESN and MIN are received, they are compared in the NTS systems Home Location Register (HLR) to validate the authenticity of the mobile radio user.

Roaming for PassPort protocol radios is automatic. The PassPort capable radio automatically communicates with its home system. As these mobiles are connected with their home system, they provide the home system database with roaming and other system information. This allows the home system to be continually updated as to the whereabouts of all radios in its talk groups. Using this information, the NTS system will connect all mobiles that are part of a specific talk group to a conversation when anyone from the talk group makes a transmission. This connection occurs regardless of what site the mobiles may be registered at in the network.

The PassPort NTS system transfers basic system information between the mobile radio and the home network and is able to connect different types of systems together. For example, 800 MHz, 900 MHz, UHF, VHF and 220 MHz trunked and conventional systems may be interconnected and controlled by a common Master Card. The systems only need to share NTS resources and software control.

Each NTS site is managed by an all digital switch. All site to site links are digital as well as all communications between the master controller and application cards at each site. The NTS network has been engineered and designed to provide for a simple migration to support future digital RF technologies such as TDMA or DCMA.

The NTS system uses dedicated hardware that is specifically designed for communications applications. This hardware is optimized for the rapid control of dispatch connections. The system is designed for redundancy. Two Master Cards may be installed simultaneously. The first (primary) card controls the system and a secondary Master Card is ready in stand-by mode in the event of failure. Should the primary Master Card fail to operate, the stand-by Master Card immediately assumes system control without any interruption. The failed assembly then notifies the system operator of the failed condition.

The PassPort trunking protocol from Trident Micro Systems is an open protocol available in radio products from multiple licensees. PassPort has been licensed by Kenwood Communications, ICOM Corporation, Ritron, Vertex/Standard, SEA, Motorola, Tait Electronics, Scholer-Johnson, ADI and ComSpace Corporation.

The PassPort trunking protocol is coordinated by NTS infrastructure. The NTS system was designed from the ground up for the trunked radio environment and is not an outgrowth of a modified telephone switch or PBX. The NTS system uses a distributed network architecture.

The distributed network allows each site to independently connect to other mobile radios and network equipment. Each site only needs to be connected to another NTS equipped site.

The NTS system allows for both LTR and PassPort trunking protocols to be used simultaneously. Each PassPort NTS network can support up to 128 sites. Networks can be combined to support up to 128 networks. In addition to the 60,000 individual user ID codes at each site, every network can allow 7.68 million roamer ID codes for temporary assignment.

When operating in the dispatch mode, the PassPort system only transmits on sites in the network with mobiles from the in-use talk group that have previously logged on or are in use during a dispatch call.

PassPort NTS radios can receive direct frequency assignments from the network. This eliminates the need for mobile radios to be programmed with every available channel they are authorized to use in the network. As new RF channels and transmitter sites are added to the network, the mobile radios can automatically be updated to find these channels.

The NTS system uses an all-digital trunking infrastructure. This removes the need for conversion to analog lines (such as E&M signaling). For more informaiton about PassPort, contact Trident Micro Systems at 828-684-7474.

APCO 16

The Association of Public Safety Officers and Communications Officials (APCO) Project 16 (APCO 16) created an industry standard for LMR systems in the late 1970s. This standard was updated to revision A (APCO 16A) in the early 1980s. This standard was primarily a User Performance Requirements wish list. No specific technical standards resulted. Each manufacturer was free to develop a solution to meet the basic requirements of the Project 16 requirements.

Project 16 systems operate on 25 kHz channels. The initial systems were analog. Some of these systems have migrated to digital with voice and data encryption supported. Various TIA/EIA documents (e.g., TIA/EIA-603) recommended specific performance measurements for LMR.

ESAS

ESAS is an enhanced LTR system that was developed by Uniden. The ESAS system allows for advanced messaging and group calls. It also allows talk around, advanced identification capability, and automatic registration that allows for wide area roaming. An adaptive scan feature automatically scans to store and prioritize an RF channel list that reflects the most often-used call groups first. This allows for a more rapid channel acquisition time.

The ESAS system allows up to 8960 ID codes per site and up to 64 service classes. The system can disable mobile radios (e.g., non-paying users) via the radio channel. The ESAS network can handle up to 1,000,000 User ID codes per network. The hub controller can be located at a network site or off-site. Up to 16 sites are addressable simultaneously. The system protocol allows for a fast 900 msec access time.

ESAS calling options include half-duplex or full duplex. Telephone interconnect and dispatch services are supported. Mobiles can be programmed for types of services including roaming, long-distance dialing, voice mail, and custom calling features. Calls can originated from standard LTR or ESAS mobiles, and 800 MHz and 450 MHz systems can be mixed.

The ESAS system continues to use the sub-audible signaling system. Because the ESAS system is an extended version of standard LTR signaling, the ESAS system can co-exist with standard LTR networks. This extended feature allows the ESAS system to provide wide-area networking, voice mail, one-to-one and group calls and other advanced features.

ESAS networks can have up to 127 sites, each having hundreds of repeaters. Mobiles use an advanced scanning system to search for the best available site signals. Once acquired, each mobile registers with a unique ID code keeping network controllers updated with their location. ESAS controllers are the main intelligence of the ESAS networked system. The ESAS system uses distributed control to minimize the effect of equipment failures. The ESAS system is managed by an advanced PC computer system that integrates the trunking controllers.

MPT 1327

Ministry of Posts and Telegraph 1327 (MPT 1327) is a standard for trunked private land mobile radio systems that are primarily used in Europe and Asia. The MPT 1327 system is well suited for small systems (even single site) that have only a few installed radio channels to large networks that may be interconnected to other networks.

The standard defines the protocol rules for communication between a trunking system controller (TSC) and mobile radio units. This MPT 1327 specification is published by the department of trade and industry (DTI) in the U.K. The original draft of MPT 1327 occurred in 1985 .

The MPT 1327 standard only defines signaling that occurs over the air. Additional standards have been developed that define performance parameters and feature operations. Other key industry standard specifications for the system include MPT 1343, which is the Air Interface specification that ensures compatibility of mobiles produced by different manufacturers. MPT 1343 performs correctly on MPT 1327 networks, thus providing true compatibility between radio units and systems originating from differing manufacturers. MPT 1347 is the radio interface specification for commercial trunked networks operating in Bank III, sub-bands 1 and 2. MPT 1352 is a performance test specification that ensures strict tolerances are met. Many other specifications that detail parts of the MPT 1327 system are available.

MPT 1327 allows up to 24 channels per site. Each trunked channel can independently operate. The system was designed to be robust so that channels can be added or removed during normal operation of the system. The system can be operated as a single site and can be upgraded for network operation at a later time. Both distributed and central processing are possible with the MPT 1327 system.

The system allows for speech calls, data calls, emergency calls, include calls and status and short data messages. Speech calls can be assigned priority levels, and various modes of speech (e.g., announcement only) are supported. Data calls can be setup that allow the regular and non-regular (signaling) transfer of data between mobile radios and the system. Emergency calls can allow the system to automatically control the mobile radio's operation. Include call allows for the addition of other users during a call. Status messages can exchange pre-defined messages between mobiles. Short data messages (about one line of text for each message) can be transferred between units.

Mobile radios are capable of initiating calls to individual radios, a group of units, all units in the system, a PABX number (up to nine digits) and a PSTN number (up to 31 digits). Status messages or short data messages may also be sent between mobiles and the trunking system controller (TSC).

When a call is set up, the TSC passes control and status information of the call to the mobile radio. This allows the system to indicate reasons for delays or failures in the system. Each mobile has individual and group IDs to which it belongs. A single mobile may be a member of many groups and the group addresses can be chosen independently of its individual address.

The system allows for up to 1,036,800 addresses per system with a maximum of 1024 channel numbers. There are 32,768 system identity codes.

The signaling data rate is 1200 bits per second. Control signaling uses Fast Frequency Shift Keying (FFSK) subcarrier modulation. A separate control channel is used. Control channels can be dedicated or non-dedicated. A dedicated control channel has a single frequency assigned for signaling. A non-dedicated system may dynamically assign the control channel for voice and data if all the other channels are in use. It is preferred to use a dedicated control for a system that has many channels.

The system continuously transmits broadcast messages to inform mobiles of key system information. This includes a list of channels that can be used for control signaling.

To help identify which channel is being used from interfering channels on the same frequency, a system identity code is used to label messages. In some cases, the channel number is also used to identify the message.

The system can ask the mobile to transmit its unique serial number at any time to determine fraudulent use. Automatic registration allows the efficient use of multi-site systems.

The control channel signaling channel is divided into slots of 106.7 msec (128 bits) each. A single message can be sent in each slot. Messages on the forward control channel (base to mobile) are normally continuous. These messages include fields, system identity and synchronization (CCSC), and an address code. The control channel system codeword (CCSC) provides the identity of the system to mobile and also provides synchronization for the message. An address codeword describes the function of the message.

The messages sent on a control channel may be random access (Aloha) requests, response requests (Ahoy) messages, acknowledgements, control messages (e.g., change channel), data, and other types (miscellaneous).

Access control on the system involves initial synchronization with the system and determination of which slots are available for access attempts. The mobile then sends a random access message during one of these slots. Because the system knows the activity of the system, it can control the number of available slots or frames for the allowing of access attempts. The mobile can inform the system of the type of call request (e.g., speech or data).

Each mobile unit address has a 20-bit number comprising two fields: a prefix (7-bits) and an identification (13-bits). Typically, all members of a group or fleet are allocated the same prefix code. Dividing the address into a prefix and unique identification code allows a message to hold two identification fields (dropping the prefix). These two identification codes can indicate the calling and called units. Extended addressing is used (multiple message words) to address multiple units with different prefixes.

The basic process for call setup from a mobile unit is a random access call setup request (RQS), instruction to send extended address information, checking availability of radio units, and finally the traffic channel is allocated and a go to channel (GTC) message is sent.

When sending data messages, the messages contain an address codeword indicating that data information will follow, and two appended data codewords will follow.

Idle mobile radios continuously identify and monitor the control channel for sync and for updates to the system information.

The most basic MPT 1327 system configuration consists of single radio channel. This radio channel will operate as a control channel until a call request is received. At that time it will reconfigure itself to traffic (voice) mode until the termination of the call.

A more typical configuration is to provide two or more channels at each site. One of the channels will operate as a control channel; the others will be traffic channels. A maximum of 24 channels are available on each site.

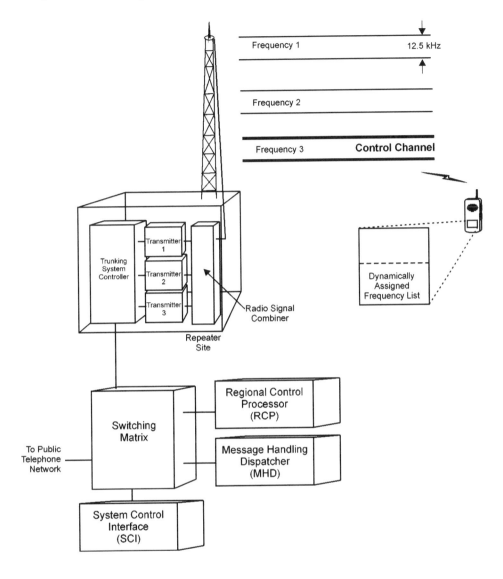

Figure 3.3, MPT 1327 System

Each repeater RF channel communicates with each other via high-speed serial data bus (SIO BUS). This allows each channel to monitor the status of other channel cards.

Multi-site systems are created by connecting each site to a central switching node. These interconnections may be dedicated private lines or dialup lines. In the event of a central switching node failure, repeaters may continue to operate in local mode.

Regional systems can be interconnected by Inter-Regional Processors (IRP) to provide very wide area coverage. Up to 16 regional systems can be linked together, providing an overall capacity of 160 sites and 3840 channels.

The basic parts of a system include the Repeater, TSC Channel Card, System Control Interface (SCI), Regional Control Processor (RCP), Switching Matrix, System Control Terminal (SYSCON), and a Message Handling Dispatcher (MHD). Figure 3.3 shows a basic MPT 1327 System.

The TSC channel card is the logic intelligence for the trunked radio functions. Each TSC controller independently controls each RF repeater. This results in a channel card for each RF channel.

A System Control Interface (SCI) links the TSC cards to the outside facilities (e.g., switching facility). This could be a regional control processor or a PC computer for a single site system. The SCI is used to transfer billing records, status reporting, system control, and mobile radio identity validation.

A regional control processor (RCP) is used to connect together up to 10 sites (SCI cards). The Regional Control (RCP) Processor provides for call setup management tasks between multiple sites. The RCP is also the gateway between the PSTN, PABXs and hard-wired dispatch consoles.

The PCM switching matrix allows for the switching of speech or data information between radio sites, the PSTN, PABXs and voice mail systems. While there are no radio call setup processing steps performed by the switch, the switch does provide for intelligence between phone systems, for example, DTMF dialing and interfacing to telephone signaling systems (e.g., E&M signaling).

System Control Terminal (SYSCON) provides for customer billing and call traffic statistics. The subscriber database is an integral part of the system control. System control also provides for the remote monitoring of equipment. Display status is provided.

Message Handling Dispatcher (MHD) is used to receive, store, process, and forward messages. An MHD has a database that contains the details of mobile radios. The MHD is connected to the RCP to facilitate gathering and delivery of messages.

Chapter 4

Digital LMR Technology

In the late 1980s, the success of wireless systems (particularly cellular systems) demonstrated the need for new types of LMR systems. As land mobile systems provide service to more customers, the systems must expand to accommodate them. If systems can expand by adding new radio sites, why change the proven analog wireless technology? The answer is twofold. First, new technologies may provide less costly ways to expand than the existing analog technologies. Second, analog technology lacks the capability to support many new services.

Digital System Advantages

To create digital systems, manufacturers either work with associations to define standards or they produce a proprietary standard that others may be allowed to produce products from. The basic objectives for digital LMR technologies include

- a significant increase in system capacity compared to existing systems;
- new features (e.g., short message services);
- improved quality service;
- added security voice privacy;
- economic benefits.

Unlike the standardized wireless systems (such as cellular and paging), early LMR systems used a variety of incompatible technologies. New LMR systems were required that could meet the long term systems service and cost objectives for LMR service providers. In the early to mid-1990s, several new digital LMR systems were developed including iDEN, EDACS, Project 25 and TETRA.

Increased Capacity

As new customers are added to an LMR system, more radio channels are needed. If the radio sites are not filled to capacity with radio channels, the LMR system operator simply adds more radio channels. If radio sites are filled or new radio channels are not available for licensing, the LMR operator may add more radio sites with smaller radio coverage areas. Either way, to expand using the old analog technologies, LMR service providers must add radio channels.

The new digital LMR technologies allow capacity increases in a different way. By allowing several customers to simultaneously share each radio channel spectrum (called multiple access) or by reducing the channel bandwidth, a carrier can increase the total number of customers they can serve with an allocated frequency band.

To upgrade an LMR system for digital service, new digital radio channels are either added or used to replace existing analog radio channels. Some digital LMR system manufacturers offer digital channels that simply replace older analog radio channel equipment.

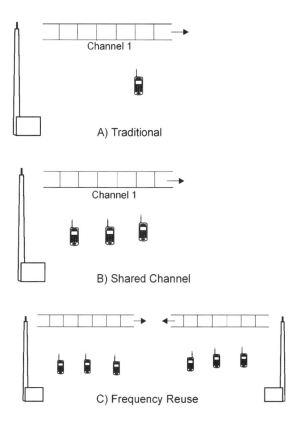

Figure 4.1, Serving More Users

LMR service providers can evaluate the potential system capacity factors of new LMR technologies by reviewing two types of efficiency: radio channel efficiency and infrastructure efficiency. Radio channel efficiency is measured by the number of conversations (voice paths) that can be assigned per the amount of frequency bandwidth (e.g., 4 users per 25 kHz). Infrastructure efficiency is measured by system network equipment and operating costs.

Figure 4.1 illustrates how a digital LMR system can allow more users to share the radio channels of a single radio site. Figure 4.1 (A) shows that an existing system is authorized to provide service on a single 25 kHz channel. In this example, a single radio channel is used as a single communication channel. This provides for one communication channel. To provide more capacity, Figure 4.1 (B) shows that a single radio channel frequency bandwidth has been divided into two radio channels that have smaller bandwidths (12.5 kHz each). This provides two communication channels. Figure 4.1 (C) shows an alternative sharing method that divides each 25 kHz radio channel into groups of three time slots (called frames). An additional method of increasing capacity through frequency reuse is where frequencies are reused at radio sites in the same network that are separated by sufficient distance so as not to cause interference.

The number of customers that can share a radio site is much greater than the number of available radio channels. This is because not everyone tries to talk at exactly the same time (except during emergencies). For each communication channel, approximately 20 to 40 two-way mobile telephone customers can share each communication channel. This allows a radio tower with six communication channels to serve approximately 120 to 240 mobile telephone customers.

New Features

Each of the new digital LMR technologies can provide similar advanced features (e.g., short message service). However, all the possible features may not be included in current or proposed industry standards for each technology. All of the new technologies allow for some advanced features that are not possible to provide with analog LMR technology, such as simultaneous voice and data transmissions. New features will bring new revenue potential. Some of the new features include

- numeric and alpha paging;
- short message transmission (text dispatch);
- priority access (for emergency services);
- group call;
- lighter and more portable units;
- longer battery life;
- vehicle location;
- imaging (video) service.

Digital Transmission Quality

As a radio signal passes through the air, distortion and noise enter the signal. A digital signal can be processed to enhance its resistance to distortion in three ways: signal regeneration, error detection and error correction. Signal regeneration removes the added distortion and noise by creating a new signal without noise from a noisy one. Error detection determines if the channel

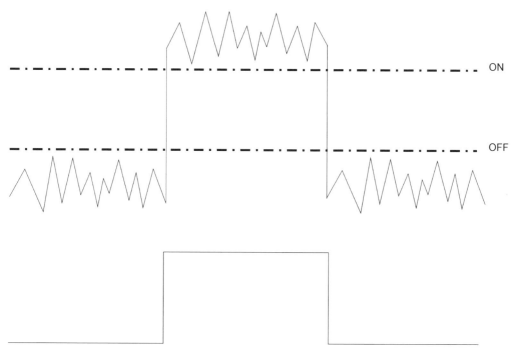

Figure 4.2, Digital Signal Regeneration

impairments have exceeded distortion tolerances. Error correction uses extra bits provided with the original signal to recreate correct bits from incorrect ones.

Figure 4.2 shows how a digital signal can be regenerated to remove the radio signal noise. In this example, Figure 4.2 (top) is a digital signal that has noise added as it has been transmitted. To remove the effects of the noise, a regeneration process is used that recreates a new digital signal using level set points. When the noisy signal crosses the ON and OFF level thresholds, a new noise free signal is created as shown in Figure 4.2 (bottom).

Security and Voice Privacy

Security involves the restriction of service use to authorized customers (called authentication) and the protection of voice signals from unauthorized listeners (encryption). Digital LMR systems use various types of authentication and encryption processes to validate users and to provide voice privacy.

Authentication is a process of exchanging and processing known information to identify specific users. To be effective, the identification information must continually change to prohibit another user from duplicating the identification information in an attempt to gain access to the system. Most digital LMR systems use an authentication process that involves the use of a software process that manipulates the stored identification information with random information (keys) that are sent on the system radio channels. Each time the mobile radio authenticates, it uses a new random number to process the identification information to produce a new authentication code. Because the system knows the key used, it can process the same information in the system to determine if the identification (authentication response) is correct.

Voice privacy is the protection of information from the possibility of unauthorized recipients listening to or decoding digital signals. Digital systems provide an immediate increase in the amount of voice privacy as most radio scanners do not have the capability of decoding digital LMR signals. In addition to the basic protection offered by digital transmission, some systems include the use of encryption programs that scramble the digital information using secret keys. These secret keys may be fixed or continuously changing. Encryption is a process of using a predetermined data manipulation process to change voice or data information. Similar to authentication, the data manipulation process must continually change to prohibit other users from listening or successfully decoding data information.

In addition to the security of voice privacy offered by system operators, the users may also add levels of security through the use of their own external authentication and encryption processes.

A key difference between LMR systems and other radio voice systems (such as cellular systems) is the requirement that several users (group users) may simultaneously listen to the same conversation. In most LMR systems, users must be capable of late entry to a conversation. This requires that voice encryption systems must be designed to allow this.

Economic Benefits

The financial goal of an LMR system is to reliably serve many users at a reasonable cost. The costs associated with an LMR system can be divided into capital (equipment) cost and operational cost.

Capital equipment cost includes radio towers and equipment, switching systems, radios and software that operates the network. Operational costs include frequency planning, maintenance and repairs, salaries for dispatchers, trainers and other costs of running the system. Digital LMR systems offer both capital equipment and operational equipment cost savings. The actual cost of digital equipment is likely to be more expensive to purchase. However, the ability to simultaneously serve several customers on a single radio channel can make the cost per average communication channel lower. The economics for digital LMR systems is discussed in Chapter 8.

System Access Technologies

Land mobile radio systems use various methods to provide service to many mobile radios using one or more radio channels. The process used by LMR systems to coordinate service requests and share radio channel usage is through the use of system access technologies. System access technologies are the language of radio and mobile radios that can only communicate with systems that can understand their language. Radio access technologies are commonly characterized by their type of access used for each subscriber. This includes frequency bandwidth, time allocation, code allocation, modulation type and types of radio channels.

There are three basic ways to divide the authorized radio frequency bands to allow radios to share communication access to the radio system: frequency division, time division and spread spectrum multiplexing.

Frequency Division Multiple Access (FDMA)

When a mobile radio uses a single radio channel during communication and when it is not in use by other radios, this is called Frequency Division Multiple Access (FDMA). FDMA was the first access technology used for two-way radios. Figure 4.3 shows how a group of radio channels can be shared by mobile radios. When a mobile radio desires to communicate with the system, it is assigned to an unoccupied radio channel. During this process, other radios are inhibited from transmitting on that communicating channel. This is usually made possible by requiring other mobile radios to sense for a RF signal strength on the channel prior to attempting to access the radio channel. After the radio channel has become available again (the mobile radio has stopped transmitting), other mobile radios can be assigned to that radio channel frequency. Mobile radios in an FDMA system may only be capable of transmitting on one frequency (fixed) or have the ability to adjust its frequency to several different radio channel frequencies.

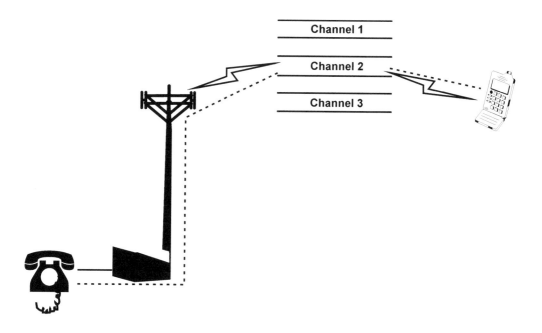

Figure 4.3, Frequency Division Multiple Access (FDMA)

Time Division Multiple Access (TDMA)

When a single radio channel is divided so that several mobile radios can use different time periods to simultaneously communicate with the system, it is called Time Division Multiple Access (TDMA). Each mobile radio that communicates with a TDMA system is assigned a specific time position on the radio channel. Because a TDMA system allows several users to use different time positions (time slots) on a single radio channel, TDMA systems increase their ability to serve multiple users with a limited number of radio channels.

To allow TDMA mobile radios to provide continuous voice communication, it is necessary for TDMA to use digital signal processing to compress digital signals into short time-slices. Figure 4.4 displays how an audio signal is digitized and compressed into a short time interval for digital radio transmission. This digital information is transmitted during a short time slot on a single radio carrier. Each mobile radio is assigned different time slots. After the compressed digital information is received, it is uncompressed and converted back into its original audio signal.

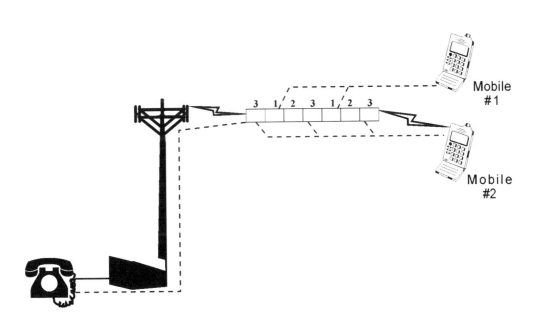

Figure 4.4, Time Division Multiple Access (TDMA)

Spread Spectrum (FHMA and CDMA)

Spread spectrum communications systems use radio channels that have a relatively wide bandwidth to allow several mobile radios to simultaneously communicate on the same RF channel. Several mobile radios may communicate on the same channel as spread spectrum mobile radios use predefined frequency hopping sequences or coded modulation codes to spread their information signal, so only small amounts of interference occur. There are two basic forms of spread spectrum communications: Frequency Hopping Multiple Access (FHMA) and Code Division Multiple Access (CDMA).

FHMA technology uses a predefined frequency hopping sequence to complete a communication path. Figure 4.5 shows how a mobile radio can communicate to a base station via frequency hopping. The time periods for the frequency hop are very small (usually only a few milliseconds). If a radio collision does occur with another user, only one or two of the frequency bursts may be lost. Because a majority of the bursts get through, the signal can be error corrected and a reliable radio channel results.

CDMA systems spread a low data rate information signal (perhaps 10 to 20 kbps) with a relatively high speed long digital code to produce a combined signal that is spread over a relatively wide frequency bandwidth. This results in many bits of code information for each bit of the information signal. When interference occurs, only a few of the coded bits are corrupted and a majority of the bits get through. Figure 4.6 shows how a single CDMA radio channel can have several communication channels through the use of different code sequences. In this example, the different code patterns that are used for each communication channel are represented by symbols. By knowing the code (symbols in this example), a CDMA receiver can use a mask (called a correlator), as shown in Figure 4.6, for each communication channel. The correlator searchers for patterns that match the programmed code and effectively screens against other codes.

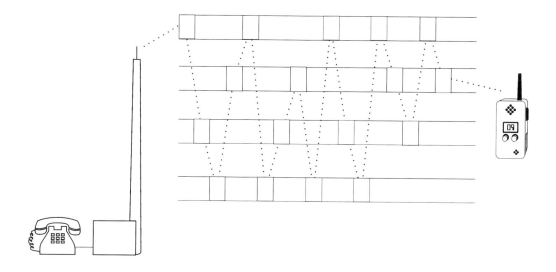

Figure 4.5, Frequency Hopping Multiple Access (FHMA)

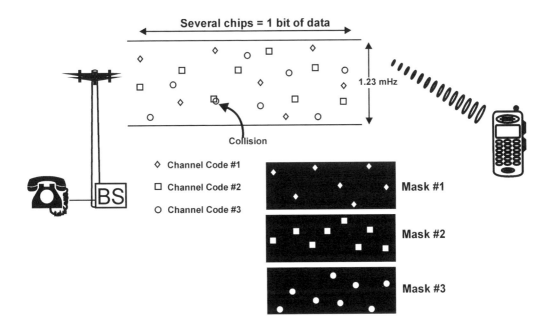

Figure 4.6, Code Division Multiple Access (CDMA)

Modulation

Modulation is the process of modifying the characteristics of a radio carrier wave (electromagnetic wave) using an information signal (such as voice or data). The characteristics that can be changed include the amplitude modulation (AM), frequency modulation (FM) or phase modulation (PM). A pure radio carrier signal carries no information. When the radio signal is modified from a normalized state, it is called a modulated signal (thus containing information). This modulated signal is the RF carrier of the information. When the radio carrier is received, its signal is compared to an unmodulated signal to reverse the process (called demodulation). This allows the extraction of the original information signal from the RF carrier.

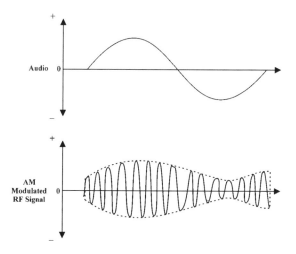

Figure 4.7, Amplitude Modulation

Amplitude Modulation (AM)

Amplitude modulation is the transferring of information onto a radio wave by varying the amplitude (intensity) of the radio carrier signal. AM is the simplest form of modulation. Figure 4.7 shows an example of an AM modulated radio signal (on bottom) where the height of the radio carrier signal is changed by using the signal amplitude or voltage of the audio signal (on top).

Frequency Modulation (FM)

In 1936, Armstrong demonstrated that an FM transmission system was much less susceptible to noise signals than AM modulation systems. Frequency modulation involves the transferring of

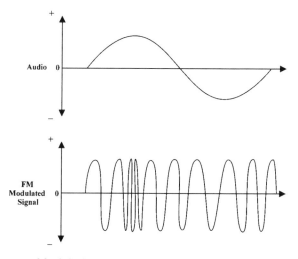

Figure 4.8, Frequency Modulation

information onto a radio wave by varying the instantaneous frequency of the radio carrier signal. Figure 4.8 illustrates the process known as frequency modulation (FM). In this diagram, as the modulation signal (audio wave) increases in voltage, the frequency of the radio carrier signal increases. As the voltage decreases, the frequency of the radio carrier signal also decreases.

One form of frequency modulation that is used to transmit digital information is called frequency shift keying (FSK). To represent a digital signal, the FSK modulator changes the transmitter frequency dependent on the level of the input signal. FSK uses different frequencies to signify a logic level of one (usually on or +5 volts) and to signify a zero (usually off or 0 volts).

Phase Modulation (PM)

Phase modulation is a modulation process where the phase of the radio carrier signal is modified by the amplitude of the information (typically audio) signal. Changes from an input source are reflected by correspondingly varying the phase (or relative timing) of the carrier wave signal as shown in Figure 4.9, which shows a sample of phase modulation (PM) with a digital input signal. In this diagram, the digital signal (on top) creates a phase modulated RF signal (on bottom). As the digital signal voltage is increased, the frequency of the radio signal changes briefly so the phase (relative timing) of the transmitted signal advances, compared to the unmodulated radio carrier signal. This results in a phase shifted signal (solid line) compared to an unmodulated reference radio signal (dashed lines). When the voltage of the digital signal is decreased, the frequency changes again, so the phase of the transmitted signal retards compared to the unmodulated radio carrier signal.

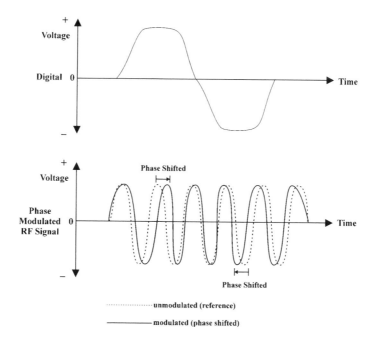

Figure 4.9, Phase Modulation

Combined Phase and Amplitude Modulation

Some of the modulation systems can combine different types of modulation (e.g., frequency, amplitude or phase) to increase the effective information transfer rate. Figure 4.10 shows the combination of phase modulation and amplitude modulation. In this example, one digital signal changes the phase and another digital signal changes the amplitude. In some commercial systems, a single digital signal is used to change both the phase and the amplitude of the RF signal. This allows a much higher data transfer rate as compared to a single modulation type.

An example of a combined modulation system is Quadrature Amplitude Modulation (QAM). QAM varies both the phase and amplitude of a RF carrier signal. The quadrature portion refers to the phase changes where each phase shift can change by a multiple of 90 degrees (+45, +135, -45, -135 degrees), and there can be several levels of amplitude. If there are four possible amplitude levels and four possible phase changes, this allows a single change (called a symbol) of a QAM signal to represent 16 possible levels (4 x 4).

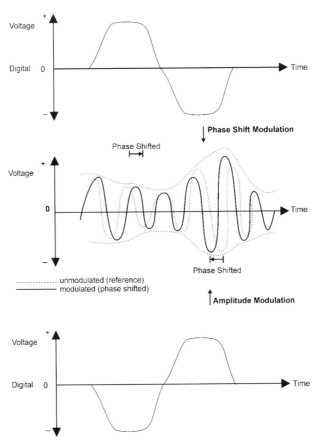

Figure 4.10, Combined Phase and Amplitude Modulation

Radio Channel Types

There are two basic types of radio channels that are used in access systems: control and communication. Control involves the coordination between transmitters and receivers so they can request service and avoid interference with each other. Communication involves the transfer of information content (voice or data) between two points. These channels may allow only one user at a time to send information or they may be divided to allow the sharing (multiplexing) of the radio channel by multiple radios at the same time.

When a radio channel is exclusively used to coordinate access to other radio channels, it is called a control channel or a signaling channel. When a radio channel is used to transfer information between radios, it is called a voice or traffic channel. Some radio access technologies can combine control and communication on the same RF channel.

Traffic (or Voice) Channels

Because digital systems use a single type of transmission (digital), individual channels are called traffic channels. Traffic channels can transfer voice or data information.

Each radio communication channel (called a "radio carrier") is typically divided into several sub-channels. The radio channel is typically called a "radio carrier" and each sub-channel is called a "communication channel." To create sub-channels, information bits that are sent on a radio channel are combined into small groups called fields. Fields are grouped into frames with pre-defined field lengths and types.

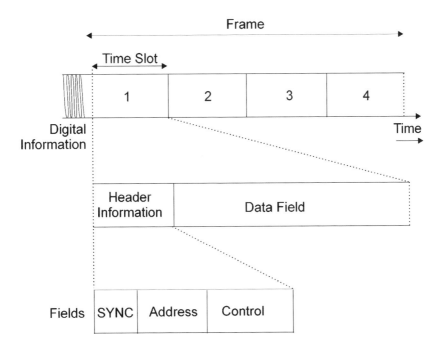

Figure 4.11, Radio Channel Fields and Frames

Figure 4.11 shows how a single radio channel is divided into fields and frames to create sub-channels. The sub-channels in this example contain system information, system status (idle/busy), paging messages and channel assignment commands. A radio listens to the radio channel to capture the information contained in one or more of the fields.

Control

Most LMR systems use a dedicated control channel to coordinate access to the system. A control channel in a digital system can be a dedicated RF carrier or a portion of a single traffic channel. The dedicated control channel in the system is dedicated to the sending and/or receiving of controlling messages between devices (typically between the base station and mobile radios). Control messages may include paging (alerting), access control (channel assignment) and system broadcast information (access parameters and system identification).

Figure 4.12 shows how control channels are typically used to coordinate access to a wireless system. In this diagram, there are two types of control channels. The dedicated control channel uses all of the information in a single radio carrier channel to coordinate access to a voice channel. The second type of digital control channel only uses a portion of a digital radio channel to coordinate access to the voice channel.

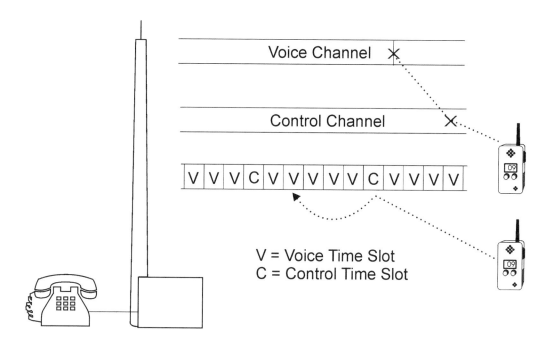

Figure 4.12, Dedicated Control Channels

Duplex Operation

Mobile radio systems typically require two-way communications where both users can communicate at the same time. To allow two-way communications, mobile systems utilize frequency division duplex (FDD), time division duplex (TDD) or a combination of FDD and TDD. TDD systems can use FDMA, TDMA or CDMA access technologies.

Frequency division duplex (FDD) systems use two frequencies to allow simultaneous communication. One frequency is used to communicate in one direction, and the other frequency is required to communicate in the opposite direction. Figure 4.13 (A) shows FDD operation.

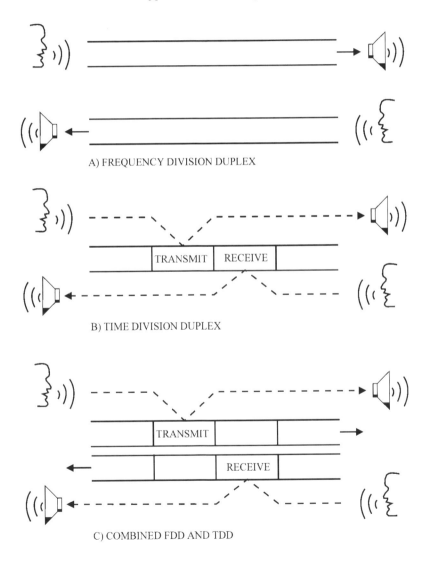

A) FREQUENCY DIVISION DUPLEX

B) TIME DIVISION DUPLEX

C) COMBINED FDD AND TDD

Figure 4.13, Duplex Systems

Time division duplex (TDD) communication uses a single frequency to provide two-way communication between devices by time-sharing. When using TDD, one device transmits (device 1) and the other device listens (device 2) for a short period of time (usually within 100 ms). After the transmission is complete, the devices reverse their role so device 1 becomes a receiver and device 2 becomes a transmitter. The process continually repeats itself so data appears to flow in both directions simultaneously. Figure 4.13 (B) shows a TDD system opera-tion. Figure 4.13 (C) shows a system that combines FDD and TDD operation. Combining FDD and TDD offers the benefit of simplified radio design.

Voice Processing/Speech Coding

Most of the new LMR systems use digital voice technology. Digital technology increases system efficiency by voice digitization, speech compression (coding), channel coding, phase modulation, and RF power control.

The first step for voice processing is to filter out unwanted frequencies from the audio sig-nal. Unwanted frequencies (high and low frequencies) can cause distortion in the digital conver-sion and compression process. The filtered signal is sampled and converted to a digital signal. The digital signal is analyzed and compressed.

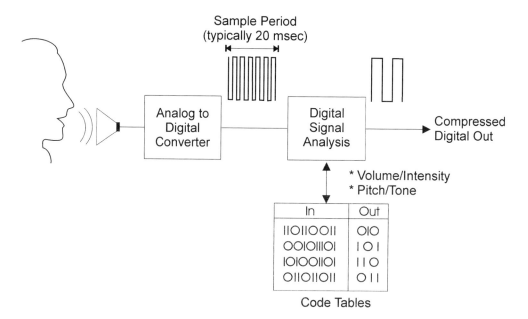

Figure 4.14, Basic Speech Coding Process

Because the digital signal is typically 64 kilobits per second (kbps), to increase efficiency, the digital signal is analyzed and compressed using a speech coder. Without speech compression, this bandwidth would make digital radio systems less efficient than analog systems, which typically use only 12.5-25 kHz of bandwidth for each radio carrier. Therefore, very high speech compression is necessary to increase cellular system capacity. Speech compression removes redundancy in the digital signal and attempts to ignore patterns that are not characteristic of the human voice. The result is a digital signal that represents the voice content, not a waveform. Figure 4.14 shows how the basic speech coding process can reduce the data rate of a digital audio signal through the use of digital speech compression (via digital signal analysis and modeling).

Channel Coding (Error Detection and Correction)

Once the digital speech information is compressed, control information bits must be added along with extra bits to protect from errors that will be introduced during radio transmission. Error protection consists of block coding and convolutional (continuous) coding. Control messages (such as power control) must be combined with speech information. Control messages are either time multiplexed (simultaneous) or they replace (blank and burst) the speech information.

Most digital radio systems use error protection systems that add approximately 50% to the total number of bits used per subscriber. Error detection/correction reduces the number of bits available to users and decreases the system capacity.

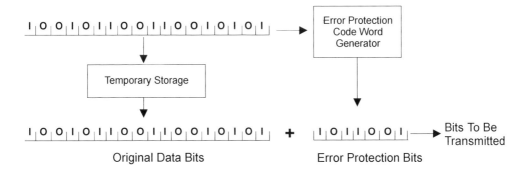

Figure 4.15, Block Error Coding

Block Coding

The block coding process adds error protection bits to the end of each frame (usually after several hundred bits) of information. These bits allow the receiver to determine if all the information has been received correctly. Figure 4.15 shows how block coding error detection and correction bits are added to the compressed speech by supplying blocks of data to a block code generator. Block CRC parity generation divides a given stimulus by a defined polynomial formula. The quotient and remainder are appended to the data stimulus to allow comparison when received. A shift register and exclusive OR gates allow the division of the polynomial.

Convolutional Coding

Convolutional coding adds bits shortly after the information is sent (usually within 5 bits). This coding allows for quick checking and correction of information during transmission in contrast to waiting for blocks of information to be received.

Convolutional coders are described by the relationship between the number of bits entering and leaving the coder. For example, a 1/2 rate convolutional coder generates two bits for every one that enters. The larger the relationship, the more redundancy and better error protection. A 1/4 rate convolutional coder has much more error protection capability than a 1/2 rate coder. Figure 4.16 shows how a convolutional coder continuously uses a continuous data signal to create a new digital signal that combines both the original information and new error protection bits.

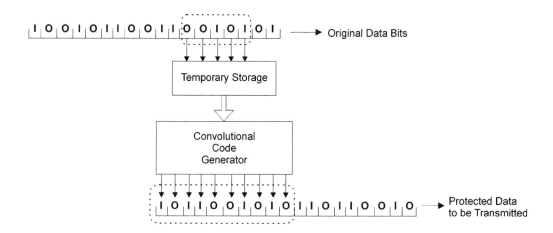

Figure 4.16, Convolutional Coding

RF Power Level Control

Most two-way land mobile radio systems use a single power level for base stations and mobile radios. Base stations typically have a high power RF amplifier to allow the signal to reach remote areas. If the range is insufficient for the low power mobile radios, repeaters may be used to extend the range.

In some LMR systems, the output power of mobile radios can be adjusted dynamically. Figure 4.17 shows that the RF power amplifier can vary its output power. In this example, the transmitted power level of the mobile radio is controlled by the receipt of messages from the base that command the mobile radio to adjust its transmitted output power level.

Figure 4.17 shows how the RF power level control process is used in most mobile radio systems. In this example, as the mobile radio moves away from the base, the base station's receiver senses a lower received RF signal level. The base station then sends a command message to the mobile radio to increase its output power level.

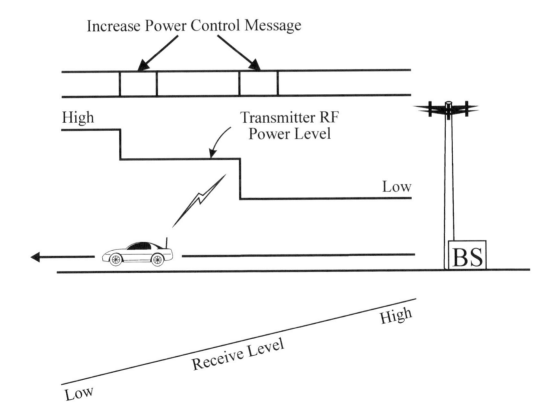

Figure 4.17, RF Power Level Control

Dynamic Time Alignment

Dynamic time alignment is a technique that allows the base station to receive digital mobile telephones' transmit bursts in an exact time slot, even though not all mobile telephones are the same distance from the base station. Time alignment keeps different digital subscribers' transmit bursts from colliding or overlapping. Dynamic time alignment is necessary because subscribers are moving, and their radio waves' arrival time at the base station depends on their changing distance from the base station. The greater the distance, the more delay in the signal's arrival time.

The base station adjusts for the delay of a transmitted signal (transmission time) by commanding mobile telephones to alter their relative transmit times based on their distance from the base station. The base station calculates the required offset from the mobile telephone's initial transmission of a shortened burst in its designated time slot. A shortened burst initial transmission is only necessary in cells where propagation time is unknown before the first transmission.

To account for the combined receive and transmit delays, the required timing offset is twice the path delay. The mobile telephone uses a received burst to determine when its burst transmission should start. Figure 4.18 illustrates how dynamic time alignment can be used to adjust a mobile radio's relative start of transmission based on the time of a received signal.

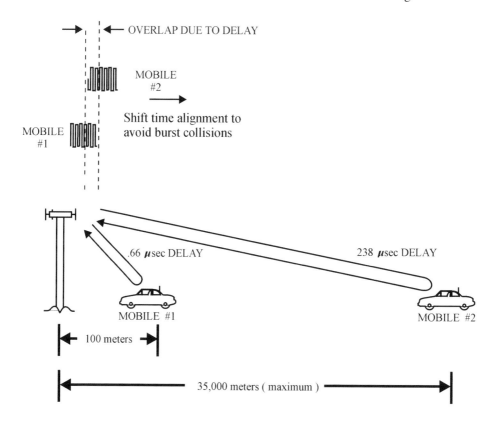

Figure 4.18, Dynamic Time Alignment

Spectral Efficiency

The spectral efficiency, or the number of users that can share a radio channel, varies depending on how the system is planned and used. The new LMR technologies have been quoted to increase the spectral efficiency by 3 to 20 times compared to existing analog systems. The increase in spectral efficiency exists despite the fact that digital transmission of a complete audio signal is less efficient than analog transmission. Digital technologies realize their gains in spectral efficiency primarily through speech coding and voice activity detection.

For digital technologies, an adequate measure of spectral efficiency is the number of bits per second per user. Digital radio technologies typically limit each user to less than 8,000 bits per second. As discussed in the speech coding section, when fewer bits per second per user are available, voice quality decreases. The basic tradeoff is increased capacity for reduced voice quality.

System Efficiency

Spectral efficiency on a single radio channel is not necessarily related to the whole system's overall cost efficiency. System efficiency is dependent upon the amount and type of equipment that is required to provide service to a number of customers. System efficiency can be measured in the number of base stations, switching centers, computers to hold the subscriber database and the leasing of communication lines between the base stations and the mobile switching center.

Each of the digital LMR technologies offers potential increased efficiencies for system equipment. The efficiency increase will be determined by both the technology (e.g., level of voice compression) and method of implementation. Base station radio transceivers become cost effective when they allow more users to share a single radio channel or when more radio channels can be installed in a single cell site. Communication lines become more efficient through voice compression that allows several users to share a single communications channel.

System Security and Privacy

Authentication is the process of validating the identity of a mobile radio to determine if it is authorized to request or receive service and, if not, to deny access to the LMR system. Authentication functions by transferring secret information between the subscriber unit and the system.

A secret algorithm is at the heart of the authentication process. The algorithm defines a mathematical manipulation of data so that if two processors have the same initial values, they produce the same answer. The answer from the authentication algorithm is used to determine if a subscriber seeking access to the system is a valid registered unit.

Authentication algorithms operate on a group of shared secret data (SSD). The SSD is contained in both the subscriber unit and system network. If either the mobile radio or LMR system have an incorrect piece of the shared secret data, authentication fails.

The key to the authentication process is not the secrecy of the algorithm that processes the information, but the initial values used when running the algorithm. When the authentication system is used, each mobile radio receives a secret number. The secret key is like the personal identification number (PIN) that is used with a bank card. The secret number is entered into the mobile radio.

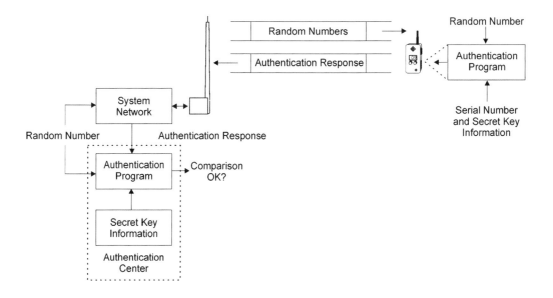

Figure 4.19, Basic Authentication Process

To add a level of secrecy, the secret key is not usually directly stored in the mobile radio. Instead, the entry of the secret key creates and stores a new secret key. After the secret key has been entered, it is known only to the subscriber and the system network operator.

Figure 4.19 shows the operation of a basic authentication process. As part of a typical authentication process, a random number that changes periodically (RAND) is sent from the base station. This number is regularly received and temporarily stored by the mobile radio. The random number is then processed with the shared secret data that has been previously stored in the mobile radio, along with other information by the subscriber to create an authentication response (AUTHR). The authentication response is sent back to the system to validate the mobile radio. The system processes the same information to create its own authentication response. If both the authentication responses match, service may be provided.

Dual Mode Systems

If a mobile radio can receive only on analog or digital technology, it could not operate in areas without its type of digital service. Users such as taxi cabs or local delivery services may find this acceptable, but others must ROAM into large geographic regions. The investment made by companies into LMR equipment increases the desire to have dual mode radios. Dual mode subscriber units can utilize one type of radio channel if available (typically digital), and if not, they can obtain service through the other (typically analog) system. An example of a dual mode radio is the combined iDEN and GSM system radio shown in Figure 4.20.

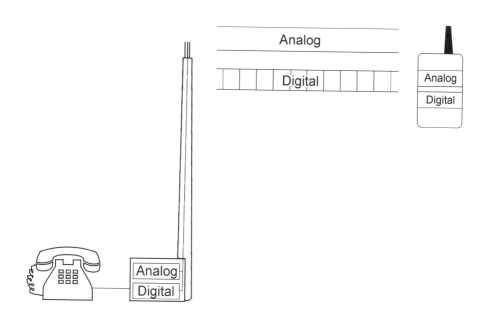

Figure 4.20, Dual Mode LMR System

Signaling

Signaling involves the transmission of unit address, supervision, or other control information between devices or switching systems. Signaling is used to establish, monitor, or release connections, provide network control and support internal operations. Signaling can be divided into two basic types: in- band signaling and out-of-band signaling.

In-Band Signaling

Also known as blank and burst signaling, in-band signaling temporarily replaces normal traffic. When signaling data is sent on the voice channel through the use of in-band signaling, audio signals are temporarily inhibited and replaced with digital messages. Messages are typically sent at low data rates of 300 to 10 kbps. Similar to the control channel signaling structure, messages are often repeated a multiple of times to ensure their correct reception.

To inform the receiver that a digital signaling message is coming, a dotting sequence (wake up alert) is sent that precedes the message. After the dotting sequence is transmitted, a synchronization word follows, which depicts the exact start of the message.

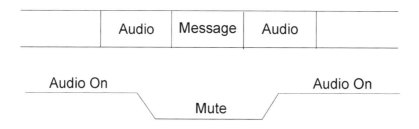

A) Analog In-Band

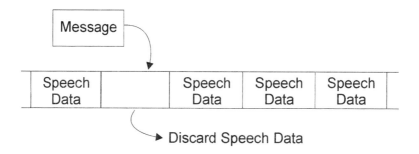

B) Digital In-Band

Figure 4.21, In-Band Signaling

Figure 4.21 shows how the basic process of in-band signaling can deliver control messages. In this diagram, the base station desires to send a message to the mobile radio. The base initially sends a dotting sequence that indicates a synchronization word and message will follow. The mobile radio detects the dotting sequence. As a result, the mobile radio mutes the audio and begins to look for a synchronization word. The synchronization word is used to determine the exact start of the message. The mobile radio receives the message, and on completion of the message, the mobile radio will then un-mute the audio and conversation continues. Because the sending of the message can be less than ¼ second, the user may not even notice a message has been received.

Out-of-Band Signaling

Out-of-band signaling involves the transfer of control information outside of the audio signal channel bandwidth. Out-of-band signaling may share a single communication channel when the communication channel is divided into several logical channels, or the signaling may occur on a separate channel or network (such as the common channel signaling on the public telephone network).

The use of out-of-band signaling allows uninterrupted communication while the users voice or data information is being transferred. Figure 4.22 shows the process of out-of-band signaling. In this illustration, two examples are shown: analog sub-band signaling and slow speed digital

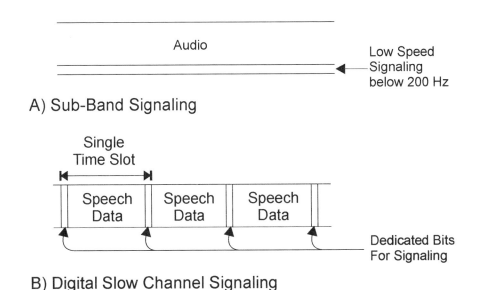

Figure 4.22, Out-of-Band Signaling

signaling. Figure 4.22 (A) shows that a low frequency control signal can be sent along with an audio signal on a radio channel. Because the low frequency control signal is below the 300 Hz minimum for audio, it is sent simultaneously with the audio signal and does not affect the audio signal. When the control signal is received at the mobile radio, audio filters separate the sub-band control signal from the audio signal. Figure 4.22 (B) shows how a portion of time slots used in a TDMA system can be dedicated for control channel signaling. Because these bits are not part of the user data bits, control messages can be sent simultaneously with the user's audio or data information.

Logical Channels

A logical channel is a portion of a physical communications channel that is used for a specific (logical) communications purpose. The physical communications channel may be divided in time, frequency or digital coding to provide for these logical channels.

Digital LMR systems typically divide their radio and intersystem communication channels into several logical channels. Figure 4.23 shows how a single communication channel is divided into several different logical channels. In this example, there are four fields per frame. Field 1 contains system information, field 2 contains the channel status information (busy/idle), field 3 transfers channel assignment commands and field 4 is defined for other sub-channels.

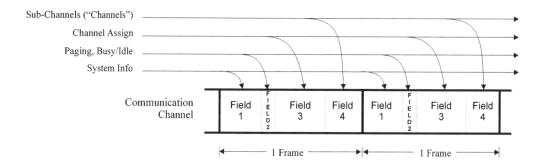

Figure 4.23, Logical Channels

Squelch Systems

Squelch systems process the audio output of a mobile radio to enable the audio output when the incoming RF signal is above a predetermined quality level. Squelch systems allow a radio user to avoid listening to noise or interference signals of distant radio transmissions that occur on the same frequency. There are three basic types of squelch: carrier controlled squelch, tone controlled squelch and digital coded squelch.

Carrier Controlled Squelch System

Carrier controlled squelch is a process where the audio output of a radio receiver is muted until the reception of an incoming RF signal is above a predetermined carrier signal strength level. Figure 4.24 shows how a carrier squelch system operates. In this diagram, the base station is receiving two signals from one desired user and one interfering user. This diagram shows that

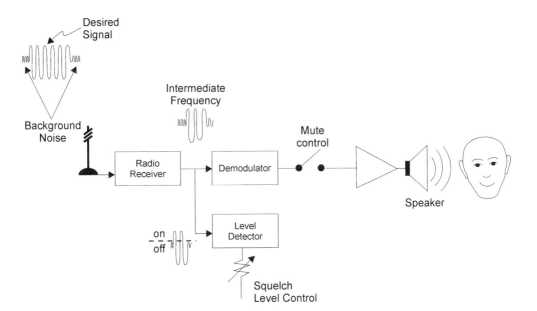

Figure 4.24, Carrier Controlled Squelch System

the squelch level has been set so the audio will be muted unless the incoming radio signal strength is above the background noise (e.g., RF signals from distant users).

Because carrier controlled squelch only senses when the incoming RF signal level exceeds a predefined level (above the background noise), when another nearby user transmits on the same frequency, conversations of other users can be heard.

Tone Controlled Squelch System

Tone controlled squelch systems mute the audio of a radio receiver unless the incoming radio signal contains a specific tone. Tone controlled squelch allows a radio user to avoid listening to noise or interference signals of other radio transmissions that occur on the same frequency from nearby users.

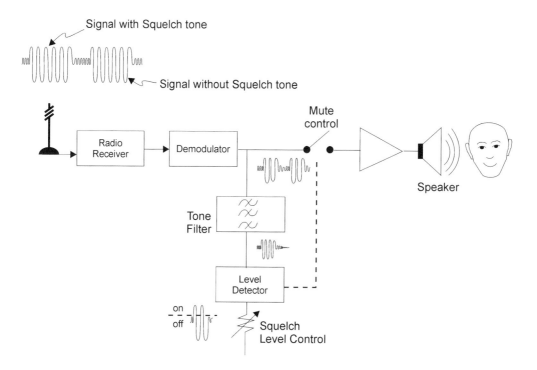

Figure 4.25, Tone Controlled Squelch System

Figure 4.25 shows a tone controlled squelch system. In this diagram, the radio receiver only un-mutes (connects) the audio signal when an incoming RF signal level contains a particular squelch tone. This permits the receiver to block out conversations from other nearby users that do not have the same squelch tone.

Digital Squelch System

Digital squelch systems mute the audio of a radio receiver unless the incoming radio signal contains a specific digital code. Digital controlled squelch allows a radio user to avoid listening to noise or interference signals of all other radio transmissions that occur on the same frequency and do not have the correct digital code mixed in with the digital audio signal.

Figure 4.26 shows how a digital squelch system operates. In this diagram, each mobile radio is assigned a unique digital squelch code. This squelch code is part of the digital message that is received. If the digital squelch code matches the prestored digital squelch code, the mobile radio will un-mute (connect) the audio path of the mobile radio.

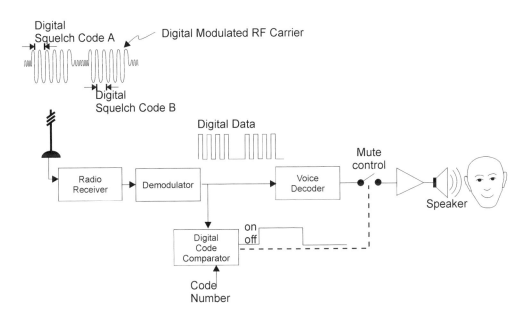

Figure 4.26, Digital Squelch System

Chapter 5

Digital LMR Systems

There are several digital systems deployed throughout the world. These include integrated digital enhanced network (iDEN®), enhanced digital access communication system (EDACS), terrestrial trunked radio (TETRA), TETRA POL and Associated Public safety Communication Officials (APCO) Project 25.

iDEN®

Integrated digital enhanced network (iDEN®) is a digital TDMA ESMR system. The iDEN® system was developed by Motorola. iDEN® is available in the United States and over 13 countries throughout the world. There are over 16 network operators that use iDEN® technology, three of which are Nextel Communications Inc., Southern Communications (Southern LINK), and ClearNET Communications (Canada). Most of the United States and Canadian population is covered by iDEN® service.

iDEN® service is used for integrated cellular-like service, paging, data and dispatch services. Key attributes include combined voice and messaging, large geographic coverage region and dispatch services.

Figure 5.1 shows a typical iDEN® system. This diagram shows that the iDEN® system communicates with mobile stations in two modes: dispatch and cellular-like. The repeater transmitters are called enhanced base transceiver system (EBTS). The EBTS contains the radio transceivers that link the radio channel to the network communication lines. The control of the EBTS is provided by base station controllers (BSCs). BSCs can control one or more EBTS units. The BSC connects communication paths to either a mobile switching center (MSC) or to a metro packet switch (MPS). The MSC can connect (switch) voice calls to the public telephone network. The MPS routes dispatch calls (called direct connect) to a dispatch applications processor.

Formerly, iDEN® was called the Motorola integrated radio system (MIRS). It was developed for the company called FleetCall, prior to it becoming Nextel. The system first offered commercial service in 1994 in California. Since its introduction, the iDEN® technology has evolved to provide higher quality voice service (cellular-like) and cost-effective dispatch voice and messaging.

The iDEN® system is a proprietary system. Developers and producers of iDEN® technology are required to have a license with Motorola. After a license is received, the air interface standards are provided. The iDEN® system uses many of the same network building blocks as the GSM system.

iDEN® is an integrated digital mobile radio system. iDEN® uses one type of digital radio channel. Each radio channel is divided into time slots to allow up to 6 users to share each radio carrier signal.

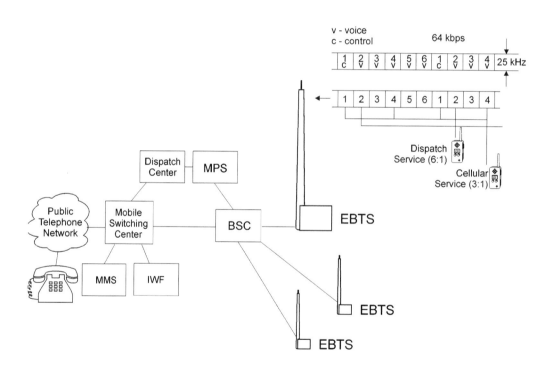

Figure 5.1, iDEN® System

There are several types of mobile radios available for the iDEN® system: portable, mobile and fixed. Portable handheld units can be divided into dispatch and cellular-like devices.

iDEN® radios only transmit using digital radio technology. However, some of the mobile radios have two modes of operation: dispatch and cellular-like. These are different voice modes of operation.

Dispatch systems use one out of six time slots for communication, while cellular-like uses two out of six time slots for communication. Dispatch mode is half duplex, while cellular-like mode is full duplex. The voice quality of the dispatch mode is lower than the voice quality of the cellular-like mode.

Control channels can be located on any authorized SMR radio channel. This complicates the initialization process for the mobile radio as it must scan many radio channels to find a valid control channel. To expedite this process, active iDEN® radio channels provide an indication of where active control channels are located within the EBTS cell coverage area.

The iDEN® system uses TDMA technology to allow up to six users to share each radio channel. The radio channel bandwidth is 25 kHz and 16 level Quadrature Amplitude Modulation (16 QAM). This allows the 25 kHz channel to provide an efficient 5.56 bits of data for each Hertz of bandwidth.

The RF power of mobile radios is 6/10ths of a Watt during the transmitter burst. The recent addition of advanced sleep modes to the iDEN® technology provides for much longer standby time battery life.

The base station RF power level is typically higher than cellular base station transmitter power levels. This allows the RF coverage area to be greater than cellular systems.

iDEN® uses vector sum linear prediction coding (VSELP) speech coding to compress the digitized audio signal into very efficient digitized voice.

The iDEN® network is composed of smart base stations (EBTS), distributed base station control (BSC), fast packet switching for dispatch operation (MPS), dispatch control center (DAP), voice switching for mobile telephone service (MSC) and advanced messaging (MMS), and interconnection to other types of systems such as the Internet (IWF).

The enhanced Base Transceiver System (EBTS) is the base station that links the mobile and portable subscribers to the iDEN® network equipment. The EBTS is the RF translating element for phone and data services.

EBTS radio transmitters and receivers are located at cell sites in order to communicate with the mobile radios. These transceivers send both control information and compressed digitized voice signals using TDMA technology. Each base radio transceiver communicates using one 25 kHz, 800 MHz channel with six time slots per frame.

EBTS radios are controlled by a base station controller. The BSC coordinates the sending and receiving of digitized voice, text messages and control messages. The BSC is the interface between the Enhanced Base Transceiver System (EBTS) and the Mobile Switching Center (MSC) that processes each type of transmission.

The BSC allows the efficient digitized voice compression to permit multiple communication channels (multiple voice channels) to share each 64 kbps communication path (DS0) between the BSC and the MSC. This allows a single T1/E1 link to carry up to four times the number of communication (traffic) channels between the switching network and the EBTS.

The Metro Packet Switch (MPS) provides the fast digital switching that is necessary to allow one-to many group calls between the Enhanced Base Transceiver System (EBTS) and the Dispatch Applications Processor (DAP). This allows for dispatch voice control.

Using the fast packet switching time, the MPS can permit wide area dispatch communications services through the rapid locating of affiliated subscriber units and rapid group call set up. The MPS allows for wide area roaming.

A Dispatch Application Processor (DAP) coordinates the overall control of dispatch communications. Because subscribers desire to send and receive dispatch calls throughout a wide coverage area, the location of subscriber units are continually tracked through periodic registration, and the information is stored at the DAP. The DAP dynamic allocates the radio site to allow the completion of dispatch calls without requiring a significant amount of setup time and inhibiting radio transmission at radio sites where group call radios are not located. This dramatically increases system efficiency.

A Mobile Switching Center (MSC) connects mobile radio calls to the PSTN or other voice network. The MSC is a switching office for all calls that are sent or received between mobile radios and the PSTN.

Software in the MSC controls the call setup and routing procedures for calls in a similar method as switching systems route and connect landline telephone calls. Unlike landline systems, the iDEN® MSC includes advanced security and validation procedures that provide access to the radio channels and other parts of the network to ensure that unauthorized use is prohibited.

The Message Mail Service (MMS) center stores and delivers text messages to and from mobile radios. Each iDEN® mobile radio has the capability to receive and store messages of up to 140 characters each. The MMS operates similar to an alphanumeric paging system.

The interworking function (IWF) adapts and buffers information between two dissimilar systems. An example of an IWF is the Internet gateway that converts email from the Internet into short messages for the MMS.

One of the most popular services for the iDEN® system is mobile telephony. Using two of the six time slots per frame, the iDEN® system can provide near toll voice quality for mobile voice. To achieve higher system efficiency (more users per radio channel), the iDEN® system only uses one time slot per six slot frame to provide dispatch service. Rapid dispatch voice connection is provided by utilizing fast packet switching technology. Late dispatch voice conversation entry to a group call is also allowed in the iDEN® system.

The iDEN® system provides for text messaging up to 140 characters per text message. iDEN® also provides for advanced battery power saving that extends battery life. The iDEN® system also has the capability to download software patches to each radio via the radio channel.

High speed packet data service and multi-mode telephones are future features. High speed packet data service (over 100 kbps) and multi-mode telephones such as combined iDEN® and GSM are planned.

EDACS

Enhanced Digital Access Communications System (EDACS) is a combined analog and digital SMR system that provides for voice, dispatch, and data trunking radio service. EDACS was developed in 1987 by Ericsson. It is used in the Americas, Asia and Eastern Europe. These systems are primarily used for (public safety, industrial, etc.) fast data and messaging access.

The present digital channel only allows one user per channel. It is anticipated that the evolution of EDACS will allow multiple users to share each radio channel via TDMA channel division. EDACS radio channels use 12.5 kHz radio channels to provide 9600 bps data. EDACS uses a dedicated control channel to coordinate system operation. The control channel operates at 9600 bps.

The common frequency bands for EDACS include 150, 450, and 800 to 900 MHz ranges. The EDACS digital channels provide for voice privacy using encryption schemes. The data channel signaling on EDACS allows for late entry of group members.

EDACS allows for automatic power level control, and mobile radios can be produced with output power up to 5 Watts. While talk around is possible with EDACS, it is not commonly used. The key attributes of EDACS include a fast radio channel access time and the ability to deploy multi-site networks (e.g., state wide systems).

Figure 5.2 shows a typical EDACS. This diagram shows that mobile radios are capable of talking to base stations (repeaters) or of direct connection with each other. The existing radio channels are all frequency division multiple access (FDMA) digital (single user per channel). This diagram shows that future versions will use time division multiple access (TDMA) technology that allows up to three users per radio channel.

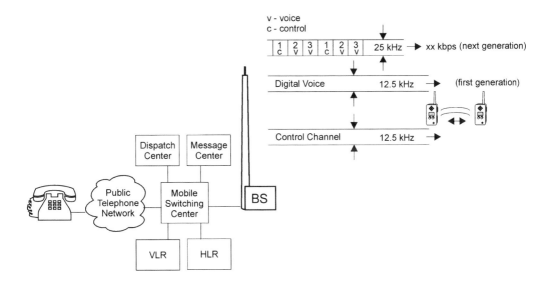

Figure 5.2, EDACS System

EDACS started from the analog systems as defined by APCO 16. EDACS evolved from the APCO 16 requirements. The standard has been proposed to the TIA, and in February 1998, a systems bulletin, PN-4137, was issued and describes EDACS. The air interface is called the digital air interface (DAI). EDACS can be used for both dispatch and telephone interconnect.

EDACS can be setup as a transmission trunking (conventional) or message trunking (trunked radio) system. EDACS radio channels are called "working channels" and a communication channel is a pair of frequencies. The base transmitter channel is called the "outbound channel" and the mobile radio transmitter channel is called the "inbound channel."

The present version of access technology is frequency division multiple access (FDMA). This will evolve to time division multiple access (TDMA) technology in the next generation. The system can have two channel bandwidths; 12.5 kHz and 25 kHz. The 12.5 kHz channel bandwidth uses Gaussian Frequency Shift Keying (GFSK) to efficiently transfer 9600 bps in a 12.5 kHz channel bandwidth.

TETRA

Terrestrial Trunked Radio (TETRA) is a digital land mobile radio system that provides integrated wireless communication services. TETRA is being developed by the European Telecommunications Standards Institute (ETSI) to create more efficient and flexible communication services from both private and public-access mobile radio users. TETRA was formerly called Trans European Trunked Radio.

TETRA began primarily in Europe as the need for public safety began to outgrow the MPT 1327 analog LMR system. Standardization of TETRA began in 1990. In 1994, it was agreed that TETRA would use TDMA technology with 25 kHz bandwidth.

TETRA is an integrated all digital radio system that is capable of sending and receiving voice and data information simultaneously. It effectively supports voice groups and has capacity for over 16 million identities per network (over 16 thousand networks per country). TETRA permits direct mode operation (talk around) that permits direct communication between mobile radios without connection through the network. TETRA includes a priority feature to help guarantee access to the network by emergency users. The system allows independent allocation of uplinks and downlinks to increase system efficiency. The signaling protocol supports sleep modes that increase the battery life in mobile radios.

The TETRA system is fully digital and allows for mixed voice and data communication. It is specified in open standards. The TETRA system allows up to four users to share each 25 kHz channel. It allows interworking with other communication networks via standard interfaces. TETRA is capable of call handoff between cells and it has integrated security (user/network authentication, air-interface encryption, end-to-end encryption).

TETRA mobile radios have the option to use a subscriber identity module (SIM) card for security keys and personal data. TETRA systems have an Inter-System Interface (ISI) that allows interconnection of TETRA networks from different manufacturers, and gateways provide access between the TETRA network and other networks. Line Station interface permits the connection of third party dispatch systems.

TETRA data rate (kbps) include unprotected, standard and high protection. Unprotected data rates are 7.2, 14.4, 21.6, 28.8, data rates for standard protection (kbps) are 4.8, 9.6, 14.4, 19.5 and data rates for high protection are 2.4, 4.8, 7.2, 9.6. The system is capable of sending pre-defined status messages (range of over 32,000 values). User messages vary in length up to 2047 bits.

TETRA operates in the VHF and UHF frequency ranges of 150MHz to 900MHz. The RF carrier spacing is 25 kHz, which is divided into four communication channels multiplexed onto each carrier (TDMA). The modulation method is phase shift keying (Pi/4 DQPSK). The channel data transmission rate of 36 kbps provides a net data throughput of 28.8 kbps (7.2 kbps per channel). The digital voice coding is ACELP.

The base station transmitter power levels include 0.6W, 1W, 1.6W, 5.5W, 4W, 6.3W, 10W, 15W, 25W and 40W, and the mobile radio is capable of 1W, 3W, 10W and 30W. TETRA base stations operate in full frequency duplex. Mobile stations may operate in frequency duplex or half duplex. The TETRA standard supports continuous, timeshared and quasi-synchronous operation.

Figure 5.3 shows a typical TETRA system. This diagram shows that TETRA radios are capable of communicating with base stations (repeaters) or each other (direct connection). The radio channels use time division multiple access (TDMA) to allow up to four users to simultaneously share a single radio channel. The TETRA system is all digital. Control channels share a single RF channel. The base station is connected to a base station controller (BSC). The BSC is connected to a Mobile Switching Center (MSC).

TETRA allows for fast call setup time for group communication and it permits direct mode communication (talk around) between mobile radios. Packet and circuit data communications are supported.

There are four phases of development of the TETRA standards. The first phase of TETRA was completed in 1995. The second phase of the standard was completed in November, 1997. As of 1999, the third phase development was still in progress, and phase four covers high bandwidth (3rd generation) technology. TETRA standards are managed by ETSI.

In phase 1 for the TETRA system, the RF carrier channels have combined voice and data using a four slot system. Phase 2 defines direct mode communication between mobile radios, design guides, validation requirements for basic service and new security algorithms. Phase 3 adds validation for direct communication services, validation of packet services, addition of a SIM card, intersystem interface (ISI) signaling for wide area roaming and beginning of phase 4 DAWS. In the future, phase 4 will allow for new frequencies that permit high speed protocol definitions.

Because the radio carrier (radio channel) is digital, applications can be developed very easily. A single radio channel can provide for shared voice, messaging and/or data services. The amount of bandwidth available to each mobile radio can be changed dynamically. The system can also operate at half or full duplex. Networks are specified in detail and can be defined at different levels (e.g., local, regional and national).

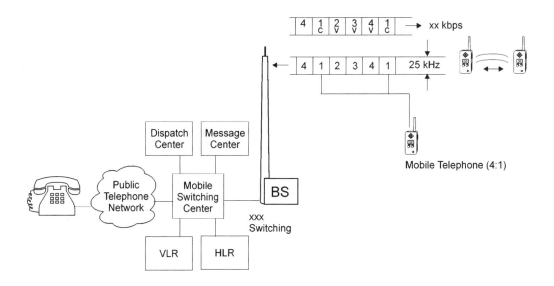

Figure 5.3, TETRA System

Basic features include individual call, group call, short data service and other supplementary services. Other advanced features include call authorization by the dispatcher (CAD), area selection (AS), access priority (AP), priority call (PC), late entry (LE), pre-emptive priority call (PPC), discreet listening (DL), ambience listening (AL) and dynamic group number assignment (DGMA).

The TETRA system uses TDMA access technology. The radio channel is divided into frames, and each frame is divided into four time slots. Each radio channel (traffic channel) is 25 kHz wide and pi/4 DQPSK modulation is used to efficiently transmit at 36,000 bps. Time slots can be combined to provide for higher data rates for each user. This could be used to increase the access speed for data services.

The control channels share one of the time slots on a radio channel. Control channels can be located on any of the authorized radio channels.

TETRA POL

Tetrapol was developed by the French company, Matra Communications, in the early 1990s and is based on frequency division multiple access technology as opposed to TETRA's time division multiple access. Tetrapol is an integrated telecommunication system for voice and data applications. Tetrapol offers cellular-like features in addition to advanced dispatch system features.

Each Tetrapol mobile radio has a unique identification code that allows individual (one-to-one) communications. This allows for telephone interconnect service. Interconnect calls can be automatic or they can be pre-screened by a dispatcher for better control of interconnect usage. Group IDs can be dynamically assigned. Late entry for group users is possible in the Tetrapol system.

The multi-site open channel control structure provides for access priorities to be assigned. Various levels of priorities can be assigned that include user priority, talk group priority, recent-user priority, preemption priority, retention priority and user-selectable call priority. With priority level assignment, pre-emptive emergency calls can be processed even when the system is fully in use. The pre-emptive call will discontinue ongoing communications of another call, if necessary. Mobile radios can also monitor the communication of several groups alternately.

Remote control of the mobile (e.g., call connect) is possible by the system operator to provide ambience listening. This feature may be used in the event no answer is received from a public safety officer. The dispatcher could activate the mobile and eavesdrop on the activity in the area to determine if a crisis situation exists.

Automatic registration is included to maintain the location of mobile radios in the system for efficient wide area dispatching and call delivery. Direct mobile-to-mobile communications (talk around) is permitted. This is accomplished by the transmission of a direct mode emergency frame on a pre-programmed channel that is monitored by all mobile radios. It is also possible for units that are communicating in direct mode (called direct mode with dual watch) to also watch the activity of the network.

The Tetrapol system uses a sophisticated authentication system to prevent unauthorized use of the system by duplicated ("cloned") mobile radios. The Tetrapol system can temporarily or permanently disable the use of mobile radios. Security keys for mobile radios can be distributed via the radio channel. This eliminates the need for key loaders. This also allows new keys to be downloaded periodically (typically daily) regardless of the size of the fleet of mobile radios belonging to a user. There is no need for the user to bring mobile radios back to be reprogammed with encryption keys. The digital channel is encrypted from end to end to protect against eavesdropping through the use of scanners. Because the encryption occurs on each end, repeater sites do not need encryption protection. Each mobile radio's ID code can be sent each time the push-to-talk activity occurs. This helps the system and dispatchers to avoid confusion and ensures that only authorized users may access the system.

The Tetrapol system uses a narrow 12.5 kHz FDMA technology (single user per channel). The modulation is Gaussian Minimum Shift Keying (GMSK). The use of GMSK permits efficient RF power amplifiers (Class C) to be used. Sophisticated Forward Error Coding (FEC) and data interleaving helps maintain audio quality even in poor radio conditions. Antenna diversity techniques can be used to increase RF sensitivity. The Tetrapol system uses a robust RP-CELP vocoder. The vocoder performs well in noisy backgrounds (e.g., yelling crowds, sirens) by suppressing background noise from the transmissions.

Because of Tetrapols ability to tolerate a high carrier-to-interference (C/I) tolerance of 15 dB, this increases the ability of the system to reuse frequencies in the system more often, increasing capacity. The Tetrapol system can use a 12 cell reuse pattern. Tetrapol also permits simulcast transmission to ensure better in-building penetration, if desired. Simulcasting allows multiple sites to transmit on the same frequency to provide inperceptible transfer for the communication path as mobile radios move between sites.

Project 25

The Associated Public safety Communication Officials (APCO) Project 25 Standard is a combination of conventional and trunked digital land mobile radio technology. Project 25 compliant systems are primarily used for public safety applications.

One of the key requirements for the Project 25 Standard was to maintain compatibility of mobile radios to operate on conventional FM channels and trunked digital radio channels. Project 25 compliant radios can operate as conventional radios with or without the use of repeaters or as digital trunked radios. This requirement allowed existing systems to gradually migrate to wide area digital trunked operation.

The Project 25 system development was started on the 15th day of April, 1992, through the creation of a memorandum of understanding (MOU) between the Associated Public safety Communications Officers, National Association of State Telecommunications Directors (NASTD), the Telecommunications Industry Association (TIA) and agencies of the Federal Government (FED). The primary purpose of the MOU was to develop a common industry standard for public safety radio users. This created the Project 25 steering committee which has become the name of the industry standard.

Project 25 technology was developed from APCO 16 (analog) land mobile radio technology. Project 25 systems are primarily deployed in North and South America.

A Project 25 system includes many standards that define system operation and network and equipment interfaces. The Project 25 standards have been submitted to American National Standards Institute (ANSI), European Telecommunications Standards Institute (ETSI), and the International Telecommunications Union (ITU) for approval. The interim standard series number for Project 25 specifications is IS-102. CCITT specifications such as ISDN and OSI were regularly used to develop Project 25 specifications.

The specifications are grouped into four key categories: high level standards and systems definitions (102), services (102.A), systems (102.B) and equipment (102.C).

There are two radio common air interfaces available for the APCO Project 25 system: conventional and digital trunked radio. The conventional radio interface is the TSB102.BAAD, APCO Project 25 Common Air Interface Operational Description for Conventional Channels. The digital radio interface standard is ANSI/TIA/EIA 102.BAAA, Project 25 FDMA Common Air Interface. In the future, a new radio interface will be defined that uses a 6.25 kHz radio channel with a more efficient quadrature amplitude modulation. The selection of the modulation technology allows for the same radio receiver design for the new technology.

The components of the RF network (sub-systems) are defined by several APCO Project 25 interfaces. These include a computer network gateway (Ed), telephone network interconnection (Et), network management (En) and Intersystem interface (ISSI).

The computer network gateway can convert network data signals to SNA, X.25, and IP to an Ethernet interface. The telephone network interface allows the radio system to be connected to standard analog telephone lines or digital trunked lines (e.g., ISDN, T1 or E1). The network management interface allows a single network management system to coordinate and control RF network equipment provided by different manufacturers. The intersystem interface allows RF subsystems from one system to be interconnected to RF subsystems of a distant system, independent of manufacturer or radio technology. For example, a Project 25 system could be interconnected to a TETRA system.

The Project 25 system also specifies interfaces to mobile radio devices. This includes the user interface and data interface (Ed). The objective of the interface standards is to allow transparent operation between accessory devices (such as a data terminal or retna scanning device) and another communication device (such as a host computer at a police station).

Perhaps the most complex interfaces are those for host computer or network connectivity. Four different types of data connectivity are spelled out in the Project 25 requirements. These include a native open interface for connecting host computers, as well as the requirement to support three different flavors of existing computer network interfacing. In practicality, this might result in only three different interfaces, if the host computer native interface is selected to be one of the other three interfaces (such as TCP/IP). The system specifications include conventional, trunked and multi-station site configurations.

The mobile subscriber unit can be a mobile radio or fixed wireless device. The capabilities of the mobile subscriber (MS) unit can include voice and/or data capability. The system was designed to easily allow for dual mode radios (analog and digital) along with two digital transmission systems (regular channels and narrowband). This provides for gradual migration from one system and technology to another.

There are three basic communication channel types supported by the APCO 25 system: 12.5 kHz digital, 6.25 kHz digital and 12.5 kHz analog (for backwards capability).

Figure 5.4 shows a simplified Project 25 system. In this diagram, there are mobile radios (called subscriber units). There are three types of radio channels used in this system: 12.5 kHz conventional, 12.5 kHz digital and 6.25 kHz digital. The radio channels can be configured as convention (single channel) or trunked (with a control channel). The Project 25 system supports the use of base stations. These base stations can be connected to a network or be independent repeaters.

The Project 25 system is a FDMA system. Radio channels can be set up as conventional or trunked radio channels. When set up as conventional, the same radio channel is used for call setup (control) and voice communications. When used as a trunked system, one of the radio channels in a base station is dedicated as a control channel and the others are used for voice communication.

The signaling protocols for conventional channels are a subset of the signaling protocols for the digital radio channels. This allows the Project 25 system to be compatible with conventional and trunked radio operation.

The APCO Project 25 system uses 12.5 kHz bandwidth radio channels. This channel bandwidth is planned to migrate to a narrower 6.25 kHz as technology evolves. This will be enabled by more efficient radio modulation technology.

The 12.5 kHz radio channel uses C4FM modulation and the 6.25 kHz radio channel uses CQPSK modulation to achieve a more efficient data transmission rate. Because of the modulation types selected, the typical receiver is capable of demodulating either the C4FM and the CQPSK signals. The maximum data transfer rate for the digital channels is 9.6 kbps.

At the start of every transmission, there is a header word. Each header word is provided with extensive error correction to ensure it is received in poor radio conditions (e.g., during a radio channel handoff). Header words are followed by voice frames that contain signaling (control information) and encryption protected user data for voice privacy.

Each header word is preceded by a synchronization word (to alert that a new header is coming) and a network identifier code. The header word includes an encipherment (encryption) type and code(s), and an destination address code (if necessary). The Project 25 system allows multiple encipherment key codes. Over the air re-keying (OTAR) of encryption keys is possible.

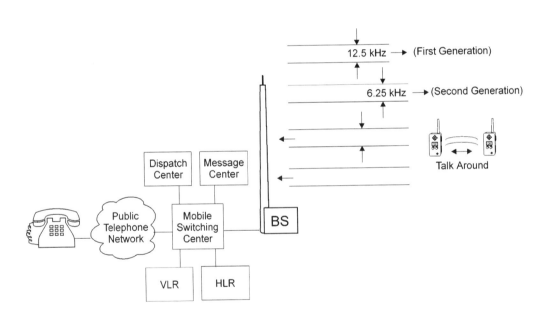

Figure 5.4, APCO 25 System

Voice frames follow the header word. Voice frames are 180 ms in length, and pairs of voice frames compose a 360 ms Superframe. Part of the voice frames are dedicated for link control. If the addressee is a talk-group, the voice frames contain information that describes the type of message to follow (group call, data, etc.), manufacturer identifier, priority indicator, talk-group address and the transmitting radio's identifier. All members of a talk group can receive the transmissions of all other members of that talk-group. Because each voice frame has a group address and type information, this allows members of a talk-group to enter the group call after it is established (late entry). The digital radio channel uses forward error correction and interleaving to allow operation when the bit error rate is up to 7 percent.

A low-speed data channel is provided in the digitized voice frame structure (88.9 bit/s). This data channel could allow applications such as accurate geographic location information or measurement of channel quality information.

Speech coding in the Project 25 system uses an improved multiband excitation (IMBE) voice coder that uses data rate of 4.4 kbps. This IMBE voice coder was also selected by INMARSAT for satellite communication devices.

There are two types of Roaming supported in the APCO Project 25 system: passive and active. In the passive mode, the subscriber unit will continuously monitor and evaluate the current control channels provided in a system without communicating with the system (no registration). When in the passive mode, information about the available control channels will be stored in the subscriber unit's memory for future use when required. In the active mode, the subscriber unit will evaluate the control channel signal quality and may attempt to be reassigned to another control channel. The subscriber unit may suggest an alternate control channel (as determined by its own control channel monitoring program) or the system may assign an alternate control channel.

The system can be set up as a simulcast or frequency reuse system. Repeaters can be connected to the network or independently operated in the system. When independently operated, the reception of the correct network ID code is used to key the repeaters. Repeaters that are connected to the network can also independently coordinate their receive and transmit channels. For example, a repeater may receive data packets when its incoming channel is idle while broadcasting a group call. The repeater automatically substitutes the incoming network ID code with a new network ID code.

It is possible for the Project 25 system to provide full duplex mobile voice services. It is also possible for the system to provide direct connection between mobile radios for voice and data transmission. The primary use of the system is for group call (dispatch) services. Even with encryption enabled during group call, late entry to the group call is allowed with minimal delay.

Messaging service is supported and message codes may be used to allow the sending of extremely small messages that represent otherwise lengthy messages.

Circuit switched data service is provided with data rates up to 9.6 kbps. The data rate is dependent on the amount of error protection used. With error protection, the bit error rate will not exceed 10^{-6}.

Packet service is supported for both mobile-to-mobile and mobile-to-network. The mobile radio can require confirmed delivery (acknowledgment) to ensure reliable transfer or send unconfirmed packets. Packet data transfer through a repeater that is connected to the network involves the temporary storage of data packets, or in the network and re-transmission of the packets when the base station transmitter channel is available.

Phase 1 mobile radios (the first generation radios) are capable of manual operation for emergency use on analog FM channels (20K0E3E and 16K0F3E) and digital radio channels (IS 102).

Priority and preemptive priority call assignment is supported. This allows emergency users with high priority to be assigned resources first and even automatically disconnect resources used by lower priority mobile radios. This is a key feature for public safety users.

Discreet listening and silent emergency features allow the mobile radio to be automatically activated for operator listening. This allows a dispatcher to monitor the activity at the mobile radio in the event of an emergency where the user cannot activate the radio (e.g., in a police shooting).

Registration is used to update the system of the region where a mobile radio is operating. There are two types of registration used in the Project 25 system: active and passive. In the active mode, the mobile radio regularly registers with the system to assist in its selection of control channels. In the passive mode, the mobile radio continually identifies the best possible control channels without registering with the system. Registration is important as it allows the system to only send messages to the radio sites where the mobile radio has recently been located. Registration is required in all systems except a single radio site system. The APCO 25 system utilizes battery saving (sleep mode) technology to extend the standby time of mobile radios.

The future features for the APCO 25 system include more efficient radio channel technology (narrowband operation) and interoperability with other systems (such as TETRA).

Chapter 6

Land Mobile Radios

Land mobile radios (LMR) convert audio or electronic signals between a person or electronic device such as a meter to a wireless system or other LMR device. Land mobile radios (commonly called "radios" or "mobiles") may be fixed in location (such as a base station) or may be mobile (such as a portable radio). Some radios may only communicate in one direction (typically a receiver) or may have two-way capability. When a mobile radio has both a transmitter and receiver contained in the same unit, it is called a transceiver.

Figure 6.1 shows a block diagram of a land mobile radio transceiver. To transmit a signal, audio sound is first converted to an electrical signal by a microphone. The audio signal is filtered, adjusted in level and sent to a modulator. The modulator modifies (modulates) an RF carrier signal using the audio signal as its basis for modulation. The modulated RF signal is then supplied to an RF amplifier which boosts (amplifies) the level of the RF signal. The amplified signal is sent to the antenna where the electrical energy is converted into electromagnetic radio waves for transmission.

To receive a signal, the process is basically reversed. The electromagnetic radio waves are received by the antenna and converted to electrical signals. Because the antenna receives other RF signals on different frequencies, only the desired frequency of the received signal is allowed to pass by a frequency filtered filter (called a tuner). This frequency may be fixed (as in the case of simple two-way radios) or it may be variable (e.g., ESMR mobile radios that can communicate on many channels). The received RF signal is then amplified and supplied to the demodulator. The demodulator uses the desired incoming reference frequency, as an example, (provided by a frequency generator) to compare the received signal with modulation to an unmodulated signal. This allows the modulator to extract the audio signal from the incoming radio frequency signal. The resultant audio signal is filtered, amplified and sent to the speaker to recreate the original audio sound.

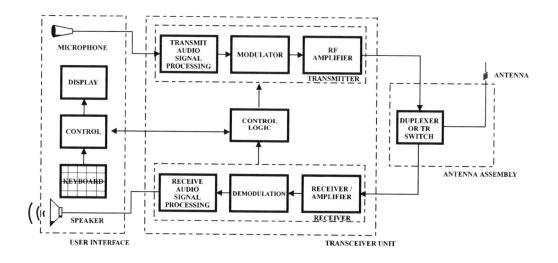

Figure 6.1, Mobile Radio Block Diagram

Analog Audio Signal Processing

Audio signals that are transmitted by analog systems (e.g., frequency modulated systems) are processed by filters, shaping circuits, signal combiners and amplifiers to change their shape and modify their content.

The first part of audio signal process is removing the frequencies that are not necessary to transmit. These unwanted frequencies are typically located below and above the audio band (300 Hz to 3300 Hz). These unwanted frequencies are blocked by a bandpass filter (signals outside the bandpass frequency band).

After the non-audio frequencies have been removed, a small portion of the audio signal may be routed to the handset speaker. This is called sidetone. Sidetone allows the user to hear a low volume of what they are saying in the ear piece of the handset.

Some LMR systems may use audio signal shaping in the form of pre-emphasis and companding. Pre-emphasis circuits shape the audio signal by providing additional amplification to the higher frequencies of the input signal. Pre-emphasis circuits are used because the signal-to-noise ratio of the higher frequencies that are applied to the modulator is lower. Companding circuits shape the audio signal by providing less gain to higher intensity (higher volume) signals. This allows people with different voice intensities (loud and quiet speakers) to have similar modulated signal levels.

When pre-emphasis and companding are used to shape an audio signal in a transmitter, their opposite functions (de-emphasis and expanding) are used in the receiver to restore the audio signal back to its original form.

In some LMR systems, additional control signal tones may be combined with audio signals prior to transmission. If control tone signals are added to an audio signal in a transmitter, they must be removed from the audio signal in the receiver by filtering. An adjustable audio amplifier is used to allow the user to control the volume of an audio.

Figure 6.2 shows typical audio signal processing for an analog radio transmitter. In this example, the audio signal is processed through a filter to remove very high and very low frequency parts. The high frequencies can be seen as rapid changes in the audio signal. After the audio signal is processed by the filter, the sharp edges (high frequency components) are removed. A portion of the audio signal is provided as sidetone to the speaker in the handset. The audio signal is then amplified and shaped by pre-emphasis and compandor circuits. At the final stage before being sent to the modulator, control tones are added to the audio signal.

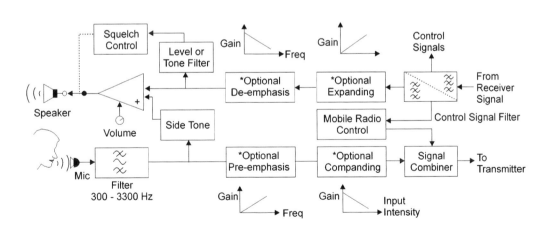

Figure 6.2, Analog Audio Signal Processing

When the audio signal is obtained from the receiver, the control signal tones are extracted by a filter assembly. The audio signal is amplified and shaped by the expandor (reverse of compandor) and de-emphasis (reverse of pre-emphasis). The audio signal level or squelch code tone of the signal is then used to control the temporary muting (squelching) of the audio signal to the speaker in the handset.

Digital Audio Signal Processing

Digital LMR systems have a very different form of audio signal processing than the processes used in analog systems. The basic difference is that digitized audio signals can be manipulated by microprocessors (limited computers) to shape them using software programs instead of electronic circuits.

The first step for digital audio signal processing is to remove the unwanted frequency components. As in analog systems, unwanted frequencies are located below 300 Hz and above 3300 Hz. These frequencies are removed by a bandpass filter. If these frequencies were allowed to pass, they could distort the analog to digital conversion process.

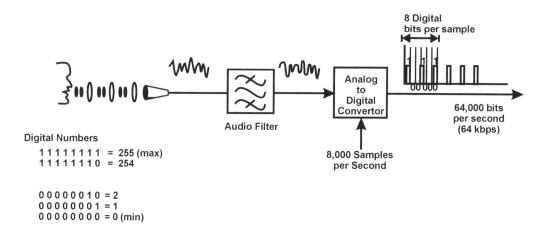

Figure 6.3, Digital Audio Signal Processing

After the audio signal has been filtered, it is applied to the analog-to-digital converter. To convert an analog signal to a digital signal, the voltage level of the analog signal is periodically sampled (usually 8,000 samples per second). This level is converted to a digital code (normally 8 bits long). The combination of 8,000 samples per second and 8 bits per sample produces a digital data signal that is 64,000 bits per second (kbps).

The digital signal is then supplied to a digital signal processor section. Early communication systems contained one or more microprocessors along with all other integrated circuits (such as memory and interface circuits) to perform digital signal processing. To reduce the number of components in a digital LMR, custom application specific integrated circuits (ASIC) and digital signal processors (DSPs) are usually used. ASICs and DSPs are signal processor circuits, containing a microprocessor, memory and stored programs, that are used to process the incoming digital signal.

Figure 6.3 illustrates the conversion from an analog signal to a digital one. Speech into the microphone creates an analog signal. An audio bandpass filter removes the high and low frequencies that interfere with digitization. The filtered signal is sampled 8,000 times per second to produce a supply of 64 kbps to a digital signal processor.

Speech Coding

To reduce the number of data bits that are transmitted, a digital speech compression process called speech coding is used. A speech coder analyzes the 64 kbps digital audio information and characterizes it by pitch, volume and other characteristics.

Figure 6.4 illustrates the basic speech compression process. The 64 kbps signal is supplied to the speech coder DSP circuit. The speech coder characterizes the input signal using the stored software program. After the digital audio signal has been analyzed, the speech coder looks up codes in a code book table which comes closest to the input signal characteristics. The code book values are then transmitted in place of the original digital audio signal. When the digital signal is received, the codes are looked up in the identical code book that is contained in the receiver. This allows the audio signal to be recreated.

The compression process may be fixed or variable. For most LMR systems, the compression ranges from 4:1 to 16:1 depending on speech activity and desired voice quality. This results in a typical compressed data rate of 4 to 16 kbps.

There are different types of speech coding processes. Because speech coders change an audio signal into a close mathematical representation of an audio signal, their quality levels vary, dependent on the mathematical model used. In general, speech coders that do not compress the signal by a large amount (high bit rate speech coders) usually have better audio quality than high compression (low bit rate). However, this varies with the type of compression process (signal analysis and selection of code book tables).

Speech coders analyze common sounds made by people. Speech coders often have a difficult time categorizing background noise as all types of background noise are not contained in the code books. This can result in audio distortion.

Another challenge with low bit rate speech coders is that they model the waveform with a small number of bits. Even when a small number of transmission errors occur, this can have a significant effect on the output waveform signals. For example, if 5 bits represent a unique sound, just a few bit errors may cause a completely different sound.

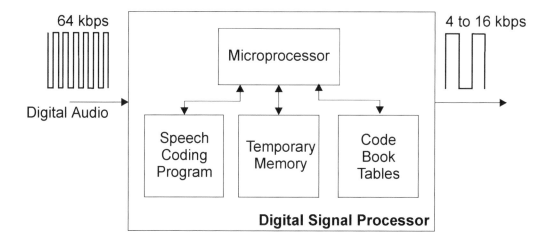

Figure 6.4, Speech Coding

The speech coding process can require a significant amount of digital signal processing. Speech coders used in digital LMR phones often require in excess of 1 to 4 million instructions per second (MIPS) to process the voice signal. It is estimated that it takes 4 times the amount of processing to increase the efficiency of speech data compression by an additional factor of two [1].

The speech compression procedure takes time to process. Speech frames are usually analyzed in 20 msec intervals. Added to this sample interval is the amount of time required for signal analysis. The combined sample period and signal analysis time can exceed 100 msec. Generally, the higher the amount of data signal compression, the more signal analysis is required. The added delay from the digital signal analysis is not usually noticeable in a two-way conversation. However, the added signal delay time can cause an annoying echo when a speakerphone is used. To reduce the effects of this echo, an echo canceling program is sometimes used to subtract the echo from the input signal. Echo canceling devices (commonly called "echo cancelers") are often installed in digital wireless networks to help reduce the effects of audio distortion.

Error Protection

Error detection and correction signal processing is used to help protect digital radio signals from signal distortion that can result in errors received from a transmitted digital signal. Error detection processing involves the sending of additional bits, with the original data signal, that are used to check if bits have been received in error. Error protection also sends additional bits along with the original data signal. However, the error protection process uses a more complex formula calculation that allows the additional data bits to check and correct some of the bits that were received in error. The error protection process is usually performed in the DSP section of the radio.

Radio Signal Processing

LMRs transmit and receive information through a radio section called a transceiver. The transmitter portion of the transceiver section modulates the radio frequently (RF) signals. Modulation is the transferring of information from an audio or other electronic signal onto an RF carrier through the proportional adjustment of amplitude, frequency or phase of a radio signal. The receiver section selects a frequency band for reception and demodulates the RF carrier signal.

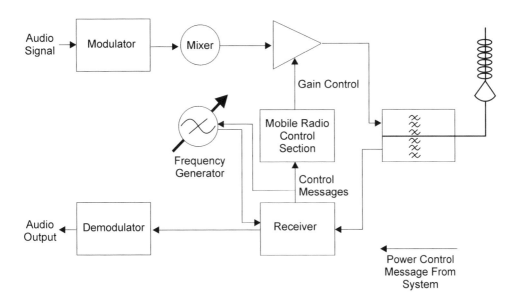

Figure 6.5, RF Power Control

Figure 6.5 shows a typical RF section of a mobile radio. In this diagram, an audio signal is applied to a RF modulator. This produces a low frequency RF signal that is converted (mixed) with a high frequency RF signal from the RF signal generator (ordinarily a frequency synthesizer). This signal is amplified and applied to the antenna. The amount of amplification for the RF power amplifier is usually fixed or under the control of the base station that it is communicating with. If the base station senses the mobile radio is moving away from the radio site, it issues a power control command to the mobile radio. This command is decoded by the receiver and the mobile radio controller increases the gain of the RF amplifier.

Display Technology

Land mobile radio display technology has changed dramatically over the past 10 years. Some LMR radios did not have any display. Today, land mobile radios have displays that can indicate the channel number and telephone number along with graphic displays that indicate battery level and signal strength level.

The trend in display technology include full graphics capability (for symbols and pictures). In the future, color graphics displays will be common. Figure 6.6 shows a typical graphics display used in a land mobile radio.

Antennas

Antennas are used to convert electrical signals to and from electromagnetic waves that travel through the air. The size and shape of antennas vary according to the frequency and type of application (fixed or mobile) and performance (e.g., gain).

The most obvious design parameter for an antenna is the length of the antenna. Antenna length changes with the frequency (wavelength) and type of antenna (directional, multi-element). The higher the frequency, the smaller the wavelength. A popular length for portable antennas is ¼ wavelength. At 300 MHz, the wavelength is one meter. This results in a ¼ wave antenna having a length of about ¼ meter (approximately 9"). The same type of antenna that operates at 600 MHz will be 1/8th meter (approximately 4.5") long.

Gain

Antennas cannot add energy (amplify) to a radio signal. However, antennas focus the energy of a radio signal in specific directions. Antennas are often characterized by the amount of signal gain in a focused direction. Antenna gain is quantified by the amount of transmitted energy that is focused in a desired direction compared to the input radio signal energy level.

Figure 6.6 shows different applications that use antenna gain. In the first example (A), a portable radio uses an antenna with limited gain of approximately 1 decibel (dB). It is necessary to use an antenna that has almost no focusing of energy. Portable radios may be used in a variety of directions and the same amount of radio signal energy should be received by the system antenna. The second example (B) shows an antenna that is mounted on a car that is traveling in a hilly area. Although the roads may have inclines and the car remains relatively horizontal with the perspective of the radio tower, it can use antenna gain (e.g., 3 dB) to extend its transmission

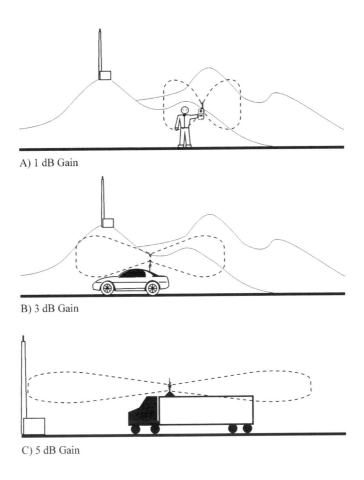

A) 1 dB Gain

B) 3 dB Gain

C) 5 dB Gain

Figure 6.6, Antenna Gain

range. The final example (C) shows a truck using a higher antenna gain in an area that has very flat terrain (such as a desert). The high antenna gain focuses the same amount of transmitter energy to allow communications to radio sites in the distance.

Antenna gain is typically expressed in decibels (dB). Decibel measurements are a logarithmic ratio of the input signal level with the output signal level. Decibel measurements are not linear. As you add dB gains, the antenna gain increases exponentially. For each 3dB of gain, this results in two times the amount of power in the focused direction. An antenna with 6 dB gain has four times the power level and an antenna with 9 dB gain has about 8 times the input power level in the focused direction.

Antenna Loaded Coils

Some LMR systems use low frequencies that would prohibit the use of ¼ or ½ wave antennas, as the antenna length (or height) would be several meters long. To overcome this challenge, low frequency antennas sometimes use loading coils. Loading coils are spools of wire that electrically extend the length of the antenna to the necessary wavelength.

Accessories

Land mobile radios can often be connected to accessories to increase the services they can provide to the user. Accessory devices include external microphones, speakers, power connectors,

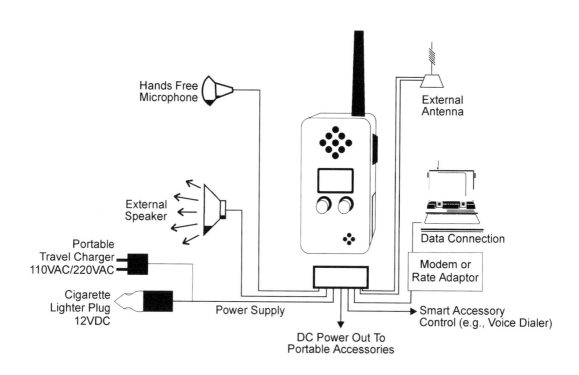

Figure 6.7, Land Mobile Radio Accessory Interface

modems, data transfer adapters, battery chargers, antennas and many other types of accessories.

To connect to most accessories, a land mobile radio usually has an accessory connector. The accessory connector may contain power input, power output, data control and antenna connections. In some cases, infrared devices can provide a communications channel to accessories (such as a computer or portable data display device). Figure 6.7 shows a land mobile radio and the typical accessories that may be connected to it.

External Microphone

It is often desirable for land mobile radios to connect to an external microphone to allow the radio to be worn on the user's belt or to be mounted on a vehicle. When the mobile radio already contains an internal microphone, the external microphone connection simply re-routes the audio path from the internal microphone to the external microphone. Some external microphones also contain a transmit button (key switch).

External Speaker

Mobile radios mounted in vehicles may have an optional external speaker. An external speaker allows the user to hear the mobile radio even when a high amount of ambient noise is present.

The external speaker is usually located near the visor; the interface circuitry connects the audio paths and allows for sensing when the user requests hands-free mode. Because digital LMR systems take time to process and convert audio signals, echoes caused by hands free operation can be very annoying to the user. To overcome the effects of echo, some land mobile radios contain an echo canceling circuit or software program to remove these echoes.

Modems and Data Adapters

Advanced LMR features and applications often involve the transfer of digital information with other devices such as laptop computers. To allow data connections, a modem or data transfer adapter must be used.

To transfer data through a radio channel, a special modem should be used. The modem should be able to adapt to the signal distortions that occur in a land radio transmission. Special error correction modems exist to increase the reliability and efficiency of data transfer when operating on a radio channel [2].

For the analog LMR system, digital information is converted by a modem to audio signals. For the digital LMR system, digital information is only buffered and shifted in time for direct transmission. When the modem signal on the analog LMR system is received by the base station, it is ready to be sent to the data device such as a computer aided dispatch station or to another device that is connected to the public switched telephone network (PSTN). When the digital information is received on the digital LMR system by the Base Station, it must be converted to a signal that can be sent to other systems through the PSTN. This is normally performed by a modem. Modems are not necessary if the PSTN has the capability to directly send digital information such as Integrated Services Digital Network (ISDN).

Battery Chargers

Battery chargers are used to recharge mobile radio batteries. There are two basic types of battery chargers: rapid charge and trickle charge. Rapid chargers supply a large amount of current to the battery to fully charge it in approximately 1 hour or less. The key limitation on the rate of rapid charging is the amount of heat generated. The more rapid the charge, the larger the amount of heat. Trickle charging slowly charges up a battery, sometimes requiring over 10 hours to fully charge the battery. Trickle charging is performed by only providing a small amount of current to charge the battery. Trickle charging may also be used to keep a charged battery at full capacity if the mobile radio is regularly connected to an external power source (such as a car's cigarette lighter socket).

The charging process (time to charge or limiting the amount of current) can be controlled by electronic circuits in the charging device (the battery charger) or by software contained in the mobile radio. The charging process involves charging a battery until a voltage transient occurs (called a knee voltage). After the knee voltage is reached, the battery charging process stops. After a battery is fully charged, some chargers will change to a trickle charge mode to keep the battery fully charged. Various chargers may fully discharge the battery first before charging. This reduces battery charge memory effect.

Software Download Transfer Equipment

Some advanced mobile radios store their operating software in re-programmable memory (flash memory). This allows upgraded software to be transferred (downloaded) after it is produced. In the past, the software installed in mobile radios required a hardware change (replacement of an integrated circuit) to update.

New operating software may allow for the installation of new features or increase the performance (e.g., RF sensitivity) of the mobile radio. New software can be transferred using a service accessory that connects the mobile radio to either a computer or data transfer box. This allows changes to be made easily in the field without opening up a mobile radio.

New digital LMR systems have the capability of updating software programs via the radio channel. The updated portion of the program is sent to the mobile radio. This update (patch) is used to change the operating program.

Subscriber Identity Cards

Some LMRs use removable identity cards. These are called subscriber identity modules (SIM). The SIM is an electronic card (chip inside a card) that holds the users identification and profile information. SIM cards are packaged in two forms: a "smart card" and a "chip carrier." The smart card is approximately the same size and thickness as a credit card. A chip carrier card

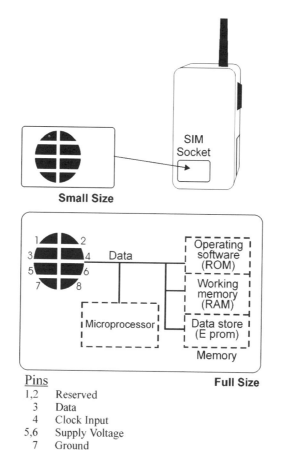

Pins

1,2	Reserved
3	Data
4	Clock Input
5,6	Supply Voltage
7	Ground

Figure 6.8, Subscriber Identity Module

is about the size of a thumbnail. Both types of cards contain the same electrical contacts that allow for the transfer of information between the mobile radio and the SIM card.

Figure 6.8 shows a block diagram of a SIM. This diagram shows that SIM cards have 8 electrical contacts. This allows for power to be applied to the electronic circuits inside the card and for data to be sent to and from the card. The card contains a microprocessor that is used to store and retrieve data. Identification information is stored in the cards protected memory that is not accessible by the customer. Additional memory is included to allow features or other information such as short messages to be stored on the card.

Power Supply

Mobile radios require various types of power to operate. The power supply may be from an automobile, line voltage (e.g., for a fixed base station) or via portable batteries.

Vehicle Power Supplies

Some mobile radios are mounted in vehicles. Automobiles typically use 12 volt power supply systems. Because most automobiles have the battery connected to the system at all times, the voltage level is fairly constant (+/- 10%). However, the manmade electronic noise from the ignition system and alternator cause voltage spikes that can be transferred via the vehicle's electrical system. These noise spikes can create audio distortion. As a result, LMR radios that are mounted in vehicles should be capable of filtering out various types of noise (e.g., ignition noise).

During some unusual conditions, the voltage of an automobile power system can change. For example, if the battery becomes disconnected or the voltage regulator fails, an alternator (voltage generator) could supply a much higher voltage. Vehicular mounted LMR radio equipment is typically designed to be protected from these high voltage spikes. However, for portable radios, the battery that normally powers radio may be the protection mechanism for these high voltage spikes. If the mobile radio is directly connected to the automotive electrical system without the battery such as with a cigarette lighter adapter that replaces the battery for a portable radio, high voltage spikes could damage the LMR radio.

In addition to using the automotive electrical systems as a power supply, some vehicular mobile radios have the capability to sense other parts of the vehicles electrical system (e.g., the ignition line). By sensing the ignition line, a LMR radio may be allowed to continue transmitting after the vehicle motor has been turned off. This is important for users that are communicating (transmitting) when the vehicle is turned off. Otherwise, the mobile radio would disconnect their call.

Line Voltage Power Supplies

Line voltage power supplies convert standard electrical signals (e.g., 110 or 220 VAC) to the DC voltages that are required by the radio equipment. Base station radios contain high power transmitters. This requires line voltage power supplies to be capable of providing high current when the transmitter is keyed on.

Batteries

LMR radios often use batteries to supply power to their systems. Battery technology continually changes. In the 1990s, new high capacity batteries were developed that overcame some of the previous disadvantages of specific types of batteries (such as memory effects).

The common types of batteries that are used in mobile radios include: Alkaline, Nickel Cadmium (NiCd), Nickel Metal Hybrid (NiMH), Lithium (Li) and Zinc Air. Batteries are often categorized as either primary or secondary batteries. Primary batteries are disposable and are not designed to be regularly recharged. Secondary are designed to be rechargeable.

Primary cells (disposable batteries) include carbon, alkaline and certain types of lithium. Because primary batteries need to be replaced after several hours of use, primary batteries see limited use in mobile radios. However, the ability to store disposable batteries for long periods of time for instant use makes primary batteries valuable for emergency conditions when recharging is not possible. It has been demonstrated that some primary batteries such as alkaline can be recharged. Recharging some types of primary batteries (e.g., specific types of Lithium batteries) can result in explosions.

Secondary batteries include NiCd, NiMH, Li and Zinc Air. Secondary batteries can be charged and discharged (cycled) a limited number of times during their useful life. The type of charge and discharge cycle can effect the number of recharge cycles. Generally, rapid charging reduces the number of recharge cycles. One of the common limitations of secondary batteries is their tendency to self-discharge. After the battery has been charged, some of the energy will dissipate within the battery. Self-discharge rates can be up to 3% discharge per day. This tends to limit the charged shelf life of the battery to only a few months.

NiCd batteries are constructed of two metal plates made of nickel and cadmium placed in a chemical solution. Because NiCd batteries have been commercially available for over 15 years, they are available in many standard cell sizes such as AAA and AA along with custom packaging sizes. Early NiCd batteries developed a memory of their charging and discharging cycles (called "memory effect"). If the charge and discharge cycles were not properly maintained, this reduced the useful life of the battery. In the mid 1990s NiCd battery design changed, resulting in a reduction or elimination of the "memory effect."

NiMH batteries provide 30% to 70% more capacity than a similarly sized NiCd battery by replacing the cadmium plate with a hydrogen adsorbing metal electrode. However, NiMH batteries tend to cost more than equivalent NiCd batteries [3].

Rechargeable Lithium batteries have been recently developed. Lithium batteries were previously only available as disposable batteries. Lithium batteries can provide 30% to 70% more capacity than NiMH batteries. A key difference between Lithium batteries and other types of batteries is that each Lithium battery provides 3.6 volts compared to the 1.2 to 1.5 volts provided by other types of batteries. The higher energy storage capacity and higher operating voltage of Lithium batteries means that a single Lithium cell can be used instead of 3 NiCd or NiMH batteries.

One of the highest capacity batteries available commercially is Zinc Air batteries. Zinc Air batteries use oxygen as one of the components to create electrical power. Zinc Air batteries have no memory effect and are low self-discharge. This allows for a long shelf life after charging. The oxygen used by Zinc Air cells is provided by the surrounding air. The by-product of the Zinc Air chemical process to produce energy is water. This requires that Zinc Air batteries have a venting area that can discharge water vapor. Another limitation of Zinc Air batteries is the limited maximum amount of current available. High current is often required when a transmitter is in operation. This limitation has been overcome by hybrid batteries that use Zinc Air to supply (recharge) another type of battery (e.g., NiMH) between transmit/receive cycles.

Digital Land Mobile Radio

Because of the advanced electronic signal processing requirements, digital land mobile radios are more complex than analog radios. The new signal processing sections of digital land mobile radios include analog to digital converters, digital signal processors, large program memory areas, phase modulators and linear RF amplifiers.

The functional block diagram of a typical digital land mobile radio is shown in Figure 6.9. The diagram shows that a handset or microphone first converts audio sound to electrical signals. These audio electrical signals are filtered to remove unwanted high and low frequencies. This filtered audio signal is then converted to a digital signal via a digital to analog converter. This produces a digital audio signal (typically a 64 kbps PCM signal).

The digital audio signal is then processed (analyzed) by a digital signal processing (DSP) section. The digital signal processing section processes digital signals using programs stored in the mobile radios program memory. Digital signal processing includes channel coding, digital speech analysis (called speech coding), error protection and other digital signal functions.

Channel coding involves the grouping of digital signals into specific frame formats (e.g., for a transmitted burst of data) and multiplexing (sharing) of control information. The digital signal processing section must also change the data rate of the incoming digital audio into the data rate of the RF channel.

The digital audio signal is analyzed by a speech coder program (algorithm). Small sections of digital audio are analyzed (usually 5 to 20 msec time intervals) to allow for data compression. LMR systems compress digital audio by a factor of 4:1 to 16:1. Data compression allows more users to share the available data on a single radio channel.

The resultant data signal is coded to include error protection and detection bits along with signaling messages. Error protection involves adding digital bits of information to the coded signal that are used to help recreate the original digital signal if radio distortion occurs during transmission that causes a loss of a few digital bits of information.

After the digital signal is processed (compressed and coded), it is supplied to a modulator that changes the digital signal (typically by phase shifting or amplitude modulating) to a low frequency RF signal. This low frequency RF signal is converted to a high frequency by mixing it with another RF signal (typically supplied from frequency synthesizer). This produces the desired frequency at a low power level.

An RF amplifier boosts (amplifies) the low level RF signal that is sent to the antenna. Because digital LMR systems often use a combination of phase and amplitude modulation, the final amplifier stage is usually a linear amplifier. The high power RF signal is passed through a frequency duplexer. A frequency duplexer is constructed of two filters in one assembly. The duplexer prohibits the high power RF signal from entering into the receiver section by permiting transmitter frequencies to only pass from the RF amplifier to the antenna and only allowing receiver frequencies to transfer from the antenna to the receiver.

The RF amplifier power level may be fixed or variable. Some advanced LMR systems automatically control the output power of mobile radios by sending power level control commands from a base station to mobile radios.

The reception process of a digital land mobile radio is basically a reversal of the process defined above. The received RF signal is initially down-converted (mixed) to a lower frequency (called the intermediate frequency). This lower frequency signal is then sent to the demodulator.

The demodulator extracts the data signal from the RF carrier signal. The demodulated digital signal is supplied to the receiving digital signal processor section. The digital signal processing section then decodes the incoming signal, extracts the control messages and decompresses the digital audio signal. The resultant digital audio signal is converted back to an audio signal by a digital to analog converter. This audio signal is amplified and provided to the speaker to recreate the original audio sound.

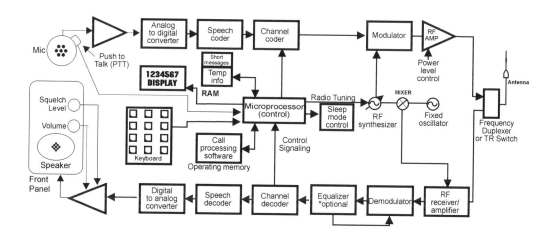

Figure 6.9, Typical Digital Land Mobile Radio Block Diagram

The electronic sections of the mobile radio are coordinated by a program that uses a microprocessor section. This microprocessor may be located inside or outside the digital signal processor section. The microprocessor section monitors the activity from the keypad (or other control device), continuously updates display (or other alert device), inserts or processes control messages when necessary and performs many other mobile radio management functions.

References:
1. Interview, Industry Expert, 22 September 1999.
2. Harry M. O'Sullivan, U.S. Patent 4,697,281, Cellular Telephone Data Communication System and Method, 1987.

Chapter 7

Land Mobile Radio Systems

The types of land mobile radio (LMR) systems vary from groups of handheld "walkie-talkies" to complex digital radio networks that have advanced messaging and data services. LMR systems can be categorized into conventional LMR, specialized mobile radio (SMR), public land mobile radio (PLMR) and enhanced specialized mobile radio (ESMR).

Conventional land mobile radio (LMR) systems are traditionally private systems that allow communication between a base and several mobile radios. Conventional LMR systems share a single frequency or use dual frequencies. When LMR systems use a single frequency where mobile radios must wait to talk, it is called a simplex system.

To allow simultaneous two-way communication or to increase system efficiency, some LMR systems use two frequencies: one for transmit and another for receive. If the radio cannot transmit and receive at the same time, the system is called half or semi-duplex. When LMR systems use two frequencies and can transmit and receive at the same time, it is called full-duplex. Full-duplex is usually reserved for telephone interconnect operation to remove the problems associated with PTT operation during a telephone conversation.

The main reason for semi-duplex operation is to allow the base station to operate as a repeater, thus effectively extending the mobile-to-mobile communication range. Signals received are demodulated and then applied to the transmitter modulator. At the same time, the received (unsquelched) signal is used to create a PTT (Push-To-Talk) signal that turns on ("keys-on") the transmitter. Mobile-to-mobile communications are thus extended to the coverage provided by the base station.

To increase the ability of a single base station to serve more mobile radios (loading capacity), sub-audible tone signaling systems are sometimes used. This included a continuous tone-coded squelch system (CTCSS) or digital coded squelch (DCS), which are employed to provide for multiple user operation. These systems use different tone frequencies or codes to uniquely iden-tify individual or groups of users. When several tones are used by a single repeater, this type of operation became known as "Shared Repeater."

LMR systems usually provide service only to a small geographic area, typically within a city or local area. The radio coverage area may even be limited to a single building or a portion of a building campus.

There are several different types of LMR systems that are owned by private companies. As a result, conventional LMR systems are not usually interconnected with each other. This limits the ability of conventional LMR radios to operate (roam) into other LMR radio coverage regions.

Figure 7.1 shows a diagram of a conventional LMR system. In a conventional LMR sys-tem, there are mobile radios (usually called "radios" or "mobiles") that communicate to other radios or a radio base (called "base"). Each radio has the capability to transmit on one radio chan-nel frequency at a time, and the mobiles have a push-to-talk button. Portable radios communi-cate by transmitting when the user presses the push-to-talk button. The LMR system is frequently coordinated by the radio users listening to radio transmissions prior to transmitting.

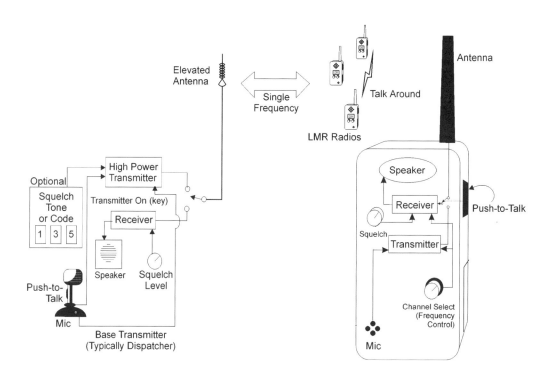

Figure 7.1, Conventional LMR System Network Block Diagram

Specialized mobile radio (SMR) systems and enhanced specialized mobile radio (ESMR) systems integrate radio technology with network system intelligence. These intelligent networks, as a rule, provide connection between mobile radios, private networks, and the public switched telephone network (PSTN).

An ESMR system is composed of radio sites, a central dispatch center and/or a switching center, customer databases and optional interconnections (gateways) to other networks (such as the public telephone network). Commonly, radio sites are interconnected to a main center via intra-system connections. The main center may be a dispatch coordinator or an automated switching center. The main center coordinates the overall allocation and routing of calls throughout the LMR system.

ESMR systems are often interconnected together (called inter-system connections) to allow mobile radios to operate in different LMR systems. This interconnection can be proprietary or standardized.

The design and layout of an ESMR system is a complex process that continually changes. This is a result of radio propagation and system loading requirements. For example, radio coverage areas can change as leaves fall off trees. The leaves on trees absorb radio energy, which

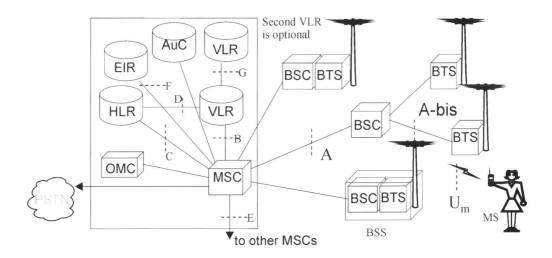

Figure 7.2, ESMR System Network Block Diagram

reduces the radio coverage area. When the leaves fall off, the radio coverage can extend and may cause interference with other radios that are operating on the same frequency. In addition to changes in radio coverage, the usage characteristics may change. For example, in the summer, more users may be at the beach or a new factory may require more radio channels to be installed.

ESMR systems provide for automated call setup, transfer and group distribution services. These services function similar to traditional cellular systems with advanced capability for group calling. Figure 7.2 shows an ESMR system functional diagram. This diagram shows that calls can be transferred between radio sites within the system (intra-system radio site handoff) or they can be transferred to radio sites in an adjacent ESMR system (inter-system radio site handoff). The overall process is coordinated by one or more switching systems. Regardless of the type of radio transmission, the switching system routes calls to and from the radio sites to other radios or to the public telephone network.

The main differences between the ESMR system and a cellular or PCS system relate to how the system architecture and signaling are designed to provide rapid call setup. ESMR systems provide rapid call setup delays in the range of 300 to 1000 ms, compared to cellular setups that can take several seconds. In addition to the rapid call setup time, ESMR systems often allow for group calling (one-to-many). This allows for dispatch type services.

Land Mobile Radio System Equipment

Land mobile radio systems are composed of base stations (radio equipment), antennas, repeaters, dispatch consoles, switching centers and databases.

Base Stations

Base station sites (commonly called repeater sites) contain the radio transmitters and receivers, controllers and communication links used to communicate with mobile radios. Base stations consist of major components like those shown in Figure 7.3. These may include one or more transmitters, receivers (for two-way systems), system controllers, communication links, and power supplies. Transmitters provide the high level RF power that is supplied to the antenna. For conventional base station transmitters, the amount of transmitter power can be over 100 watts. For ESMR systems, the amount of transmitter power per radio channel rarely exceeds 50 watts. The maximum ERP (Effective Radiated Power) is usually restricted by the license issued from the local regulatory body.

Base station receivers select, amplify, and demodulate incoming RF signals from mobile radios. Many ESMR systems use two antennas and two receivers to allow diversity reception. Diversity radio reception uses two antennas to receive the same RF signal to minimize the effects of radio signal fading. The antennas are separated by enough distance (normally a few meters when used in a mobile communications system) so a radio signal experiencing a radio signal fade, may continue to be stronger on the other antenna. The diversity reception system either combines the signals from the two antennas or selects the strongest received signal.

Controllers coordinate the overall operation of the base station equipment and provide alarm monitoring of electronic assemblies. Communication links allow a controller (such as a dispatch console or switching system) to control and exchange information with the base station equipment.

Base station radio equipment requires continuous power supply. Most base stations contain primary and backup power regulators and supplies. A battery typically maintains operation when primary power is interrupted. A generator may also be included to allow operation during extended power outages.

Figure 7.3 shows a conventional radio base station block diagram. In this diagram, the front panel of the radio base station contains a speaker and level controls that direct the base station operation. In addition, a microphone with a transmit (key) button is also connected to the base transceiver front panel.

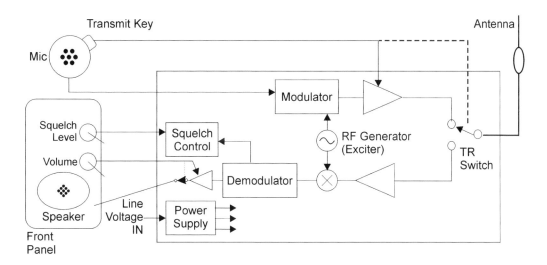

Figure 7.3, Conventional LMR Base Station Block Diagram

The microphone converts the dispatchers voice to electrical signals that are supplied to the modulator. The modulator modifies (modulates) a low level RF signal which is supplied to the mixer. The mixer combines (multiplies) the low frequency modulated RF signal with a reference frequency signal to a high frequency signal. Older equipment multiplied the reference signal by the final frequency (commonly called an "exciter"). Most present day equipment employs a programmable synthesizer, locked to a reference oscillator, to produce the exciter output. The RF frequency is then amplified by a high power RF amplifier (possibly to 100 watts or more). The RF amplifier is turned on when the transmit (key) button is depressed by the dispatcher. The RF output is routed through the transmit-receive (TR) switch. The TR switch connects the antenna to either the transmitter or receiver. This ensures the high power of the transmitter will not be connected to the receiver. If such a connection were to occur, it is likely that the sensitive receiver circuits would be damaged.

Radio signals that are sent by mobile radios in the field are received by the antenna, passed through the transmit-receive (TR) switch and converted down to an intermediate radio frequency by a mixer. This intermediate frequency is converted back to its original audio signal by a demodulator. The audio output is amplified (controlled by the volume control) and sent to the speaker. A squelch level control is used to disconnect the audio signal if no radio activity is detected (squelch control).

For ESMR systems, each base station site generally has several installed radio transmitters and receivers. One of these radio channels or a portion of a digital radio channel is used as a control channel. The control channel sends system information and paging messages and provides an access channel for mobile radios attempting to use the system.

Base stations for ESMR systems are usually equipped with a radio channel scanning receiver (called a scanning or locating receiver), which measures the signal strength and channel quality during handoff call processing. One of the advantages of digital systems is the ability of a mobile radio to monitor its received radio channel quality to provide this information back to the LMR system. This can assist the mobile switching center (MSC) in the transfer of calls between base stations (called handoff).

Figure 7.4 shows a typical ESMR base station block diagram. In this example, there are two radio channel transceivers. Each transceiver contains a transmitter and two receivers (for diversity reception). Most of the transceivers are used for voice communication. One is dedicated for control channel use (for paging and access coordination). A locating (or scanning receiver) is included to allow the sensing of the received signal strength from mobile radios to assist in the handoff process. A controller coordinates the overall operation of the base station. An incoming communications line (T1 or E1) contains multiple communications channels (64 kbps each). One of these communications channels is used for control and the others are available to carry voice communications between a switching system and the transceiver.

Each radio transceiver is connected to a high power RF amplifier. The high power RF amplifier typically boosts the RF signal strength from a few milliwatts to over 50 watts.

Each RF amplifier is connected to a RF combiner. The RF combiner allows all the transmitters in a base station to share a common transmitter antenna. The RF combiner consists of RF bandpass filters (ordinarily tuned cavities) that allow the single frequency of one transmitter to pass to the antenna but reject the frequency of other transceivers installed in the base station. The combiner also provides isolation between transmitters to reduce the generation of spurious (intermodulation) signals that can cause interference to the system or other system receivers.

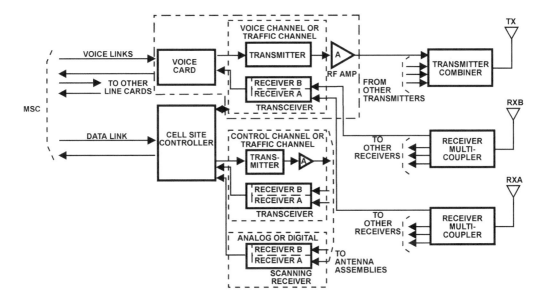

Figure 7.4, ESMR Base Station Block Diagram

Each receiver antenna is connected to a RF multi-coupler. The RF multi-coupler splits (and commonly amplifies to recover splitting losses) the incoming RF signal so it may be connected to multiple receivers in the base station. RF multi-couplers may also contain RF amplifiers to overcome the losses due to signal splitting.

Antenna Towers

Antenna towers raise one or more antennas to a height that increases the range of a transmitted signal. Antennas may be mounted on poles, guided towers or free standing constructed grids, and these towers can vary in height from a few feet to more than 300 feet. A single radio tower may host several antenna systems that include paging, microwave or cellular systems.

To provide radio signal coverage to specific areas, radio towers are strategically located around a city. Control rooms are located at the base of the towers and contain the components to operate the radio portion of the communications system.

Radio towers raise the height of antennas to provide coverage to a specific area. Radio site antenna heights vary from approximately 20 feet to over 1000 feet. When mounted on a radio tower, the antenna system can be shared with other system equipment such as a paging system antenna. Radio towers sometimes support microwave antennas that link radio sites together.

The different types of radio towers include monopole, guided wire, lattice and man-made structures. Monopole heights typically range from 30-70 feet. Guided wire tower heights can exceed 300 feet. Lattice (free standing) towers range from approximately 20-100 feet. Man-made structures include water towers, office buildings, and church steeples. Some radio site antennas are disguised to fit in with the surroundings.

Figure 7.5 shows various types of radio towers. A free standing single pole antenna called a monopole is shown in Figure A. Figure B shows a guided wire tower. A free standing antenna is a three or four-legged structure that supports itself (Figure C). System antennas are sometimes located on building tops to focus their radio energy to specific areas. Antennas can also be disguised by placing them inside a building.

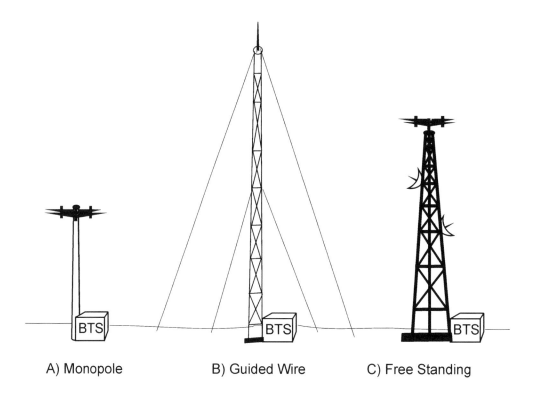

A) Monopole   B) Guided Wire   C) Free Standing

Figure 7.5, Types of Antenna Towers

Repeater Relays

Repeater relays (often called repeaters) receive a radio signal from a nearby radio site, amplify that signal and re-transmit it in a new direction. The use of repeaters plays a vital part in providing cost effective LMR service to large geographic areas.

Some repeaters receive, amplify and re-transmit the same frequency while others receive, decode and re-transmit on a new frequency. Under certain conditions, it is possible to extend the range of analog and digital channels by the use of repeaters.

While it is possible to extend the range over a hundred miles, this increased range may introduce propagation delays and radio signal (amplitude and phase) distortion. The propagation delay introduced by repeater amplifiers may exceed the maximum delay offset the mobile radios are capable of adjusting to when operating on a digital traffic channel. Propagation delays will also cause simulcast effects if there are coverage overlap areas between the donor and repeater sites. In addition, existing repeaters may use class C amplifiers that introduce phase distortion, which may result in high or unusable bit error rates (BERs) when repeat amplifying a digital radio channel.

Dispatch Consoles

Dispatch consoles allow an operator or computer to provide phone connections or to manage the coordination of communication to groups of users. Phone patching allows a radio circuit to be temporarily connected to a public telephone network. Dispatch consoles can be simple patch panels to highly automated computerized communication systems that can coordinate and/or track mobile vehicles.

The dispatch console can be coordinated with other consoles or can be autonomous. This land mobile communication system and console allows a dispatcher to select communications with individual mobile radios or to select communications with some groups or all mobile radios. The dispatcher can also cross patch communications between mobile radios and the phone system.

Advanced dispatch systems are called computer-aided dispatch (CAD) systems. CAD systems can be automated messaging devices to complex computer systems that display maps and vehicle positions on a computer monitor. The CAD console tracks the position of vehicles and automatically routes messages between groups of mobile radios.

Switching Centers

Switching centers are normally used in two-way mobile communication systems to allow the connection of mobile radios to other radios in the system or to the public telephone network. When used in an ESMR system, the switching system is usually called a Mobile Switching Center (MSC). The MSC, just like a local telephone company, processes requests for service from a customer (mobile radio) and routes the calls to other destinations.

Figure 7.6 illustrates a wireless switching system's basic functional components: a customer database, switching system, communication controllers, primary and backup (batteries) power supplies and an interface to the radio base stations (BS).

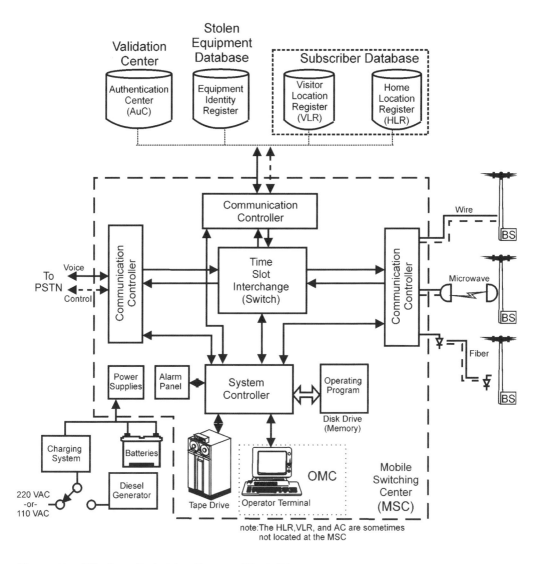

Figure 7.6, Wireless Switching System Block Diagram

Databases

Customer databases are computer storage devices (normally a computer hard disk) that hold the identification and feature preferences for each customer registered in the system. Each wireless customer (subscriber) has a profile in the main system database. This database is ordinarily called the Home Location Register (HLR). The HLR identifies the current system location of the mobile radio, the most likely place for the mobile to be or the last location the subscriber was active. The switching system uses this information to route calls to the appropriate radio tower for call completion.

The main database (HLR) holds all the detailed customer profile information while a temporary database called the visitor location register (VLR) holds information on the active customers in the system. This includes their present location within the system and registration information. VLRs hold both active home customers and visiting users from other systems.

When a new customer enters into the system service area, the mobile radio automatically attempts to register with the system (called autonomous registration). To ensure the mobile radio is authorized for service, the system will first check the VLR to see if the user is currently registered with the system. If it does not find information in the VLR, the system will use the registration request information (phone number) to determine the location of the user's HLR. The system will then send a request for the user's profile to the database (HLR) that services the mobile radio's home system. If the request returns a valid user, a temporary registration will be established in the VLR and the new user may begin to obtain service from the system.

System Security

In some wireless LMR networks, allowing access to system services requires validation of the mobile radio's identity. These systems may use an Authentication Center (AUC) to store and process secret data to stop fraudulent calls or prohibit access to other paid-for subscription services.

When mobile radios request service from the system, they transmit some of their identification information over the radio airwaves when they attempt to access the system. Thieves are able to receive this information and may try and duplicate (clone) the identification information. This could allow them to make calls that would be billed to other radios if no authentication measures were in place. To prevent unauthorized duplication of identification information, a security system (called "authentication") can be used. The authentication system uses secret keys to constantly change the identification information each time an access occurs. If a thief captures and duplicates the previous identification information, access will not be granted because a new code is required.

The authentication process uses coded keys that are created from secret codes that are stored in both the mobile radio and in the system. These keys are sent along with the basic identification information (e.g., mobile radio phone number) that is transferred during each access attempt to obtain service from the system. Only the coded keys are transmitted, not the secret keys. Since both the system and the mobile radio contain the same secret keys to create the coded keys, only the mobile phone and the system can validate if the coded key information is correct. If the coded keys do not match, the system can keep the call from being processed or disconnect the call while it is in progress. During each new access attempt, a new code is created. This prevents the copying of the secret keys themselves.

Implementation Methods

LMR systems have various design options. These options include radio simulcast transmission, frequencies reuse, overlay or integrated, sub-rate multiplexing and decentralized (distributed) switching.

Radio Simulcasting

To provide continuous radio coverage throughout a wide geographic area, some LMR systems use simulcast radio transmission. Simulcast radio transmission involves the use of multiple radio transmitters that operate on the same frequency in a geographic area. Benefits of simulcast transmission include better in-building penetration, continuous radio coverage of large geographic regions, and simplified, low cost single frequency receivers.

Two transmitters that are simulcasting a radio signal on the same frequency are shown in Figure 7.7. Simulcasting allows the radio coverage from adjacent radio transmitters to overlap. Because the transmission time from each of the transmitted signals may not be the same (one of the transmitter towers may be closer to the mobile radio than the other), the information sent by adjacent transmitters may be synchronized in time so that it arrives at the same approximate amplitude and phase as the other transmitter. Otherwise, this may cause radio signal distortion that could cause transmission errors. This is especially important in radio transmission systems that operate at high data rates (e.g., 9600 bps compared to 1200 bps).

To help combat the effects of combining slightly delayed (out-of-phase) radio signals that can cause dead spots (signal fades), the radio frequency of the various base transmitters may be intentionally adjusted so each have a slightly different frequency. With a difference of only 50 to 200 Hz, this may reduce the effects of radio signal fading.

Because simulcast transmission sends the same information to each transmitter, there is no increase in system capacity through the addition of radio towers. This results in a higher average infrastructure cost for each customer. When a simulcast system is operating near its service capacity, adding another simulcast transmitter only increases the radio coverage area, not the ability for the system to service more customers (see Chapter 8).

One-way paging systems often use simulcast systems. Providing the same paging information via multiple transmitters helps ensure that pages are received throughout a wide geographic area.

Offset Radio Channel

Offset radio channel frequency assignments can be used to reduce interference to nearby transmitter sites that are operating near the same radio frequency. When radio channels are in use, the radio energy is distributed toward the middle of the frequency bandwidth. While there remains interference to nearby transmitters that are operating on the offset frequency, the interference level is greatly reduced. The use of offset radio channels increases the ability to reuse frequencies in a wireless system, thus increasing its maximum system service capacity.

Frequency offsetting is in use in the 450 to 470 MHz frequency bands available in the United States. The primary channels for these systems often use 25 kHz channels, and offset channels are spaced 12.5 kHz from the primary channels. With 302 primary 25 kHz channel pairs, this provides up to 615 offset channels.

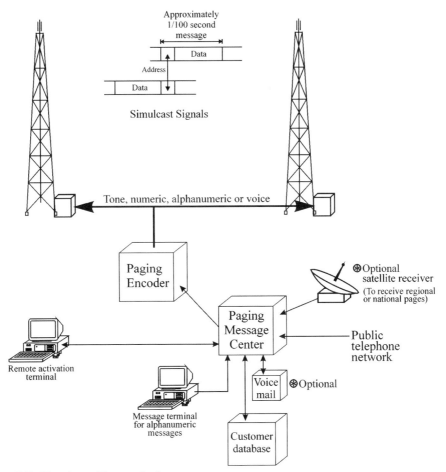

Figure 7.7, Simulcast Transmission

Frequency Reuse

Early LMR systems used a single high power transmitter to provide service to a specific geographic area (e.g., city suburb). The frequency assigned to a high power transmitter would normally not be used in other nearby cities as interference would occur. Since there were a limited number of radio channel frequencies available, this limited the total number of radio channel frequencies that could be used in a specific geographic area. Also, there were only a few radio channels available for mobile service, and this dramatically limited the number of mobile radio customers that could be serviced.

Late in the 1970s, it was demonstrated that the number of communications channels within a geographic area could be dramatically increased through coordinated frequency reuse.

By using low power radio transmitters, the same frequency can be used within a short distance. This allows many radio transmitters to be installed in a specific geographic territory. Because frequency reuse allows different mobile radios to transmit and receive on the same radio channel frequency without interference, many more mobile radios can be simultaneously serviced within a limited geographic region.

Figure 7.8 shows a frequency reuse system that has four radio transmitter sites. The radio frequency at transmitter site #1 is reused at transmitter site #4. The radio signal interference level from transmitter site #4 is usually less than 1% (20 dB below). This is called co-channel interference. Mobile radios can typically tolerate interference levels of 2% to 4% (17 dB to 14 dB below) without significant received signal distortion.

Frequency reuse is used in most ESMR or public LMR systems. Frequency reuse is usually planned by radio system designers. Frequencies are typically assigned to each transmitter site so that adjacent and alternate radio sites do not operate on the same radio channel frequency. The frequency assignment process can be very complicated. Because radio propagation can vary due to a variety of factors, a frequency plan is usually started by using system frequency planning software. This software estimates radio propagation factors and assigns frequencies so they do not exceed a desired interference level. Because all the radio propagation factors cannot be accounted for in the software program (e.g., exactly how much radio energy will be absorbed by tree leaves), temporary radio towers are sometimes used to obtain real radio signal propagation measurements. These field tested levels help the frequency planning process.

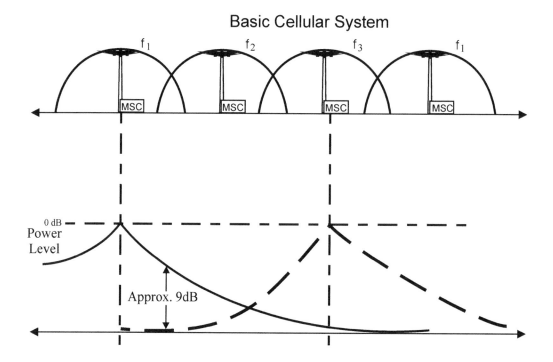

Figure 7.8, Frequency Reuse

As additional transmitter sites are added to an ESMR system (due to system expansion or radio coverage requirements), transmitter site frequencies and power levels must be re-adjusted. For most wireless systems that are growing, this is a continual process.

The minimum acceptable distance between transmitter sites that can use the frequency are determined by a distance to reuse (D/R) ratio formula. The D/R formula is a constant ratio of the distance between transmitter sites that use the same radio frequencies (D) to the radius (R) of the larger transmitter site radio coverage area. A D/R ratio for an FM LMR system is approximately 4.6 times. For example, a transmitter site that has a one kilometer radius would not interfere with another transmitter site that is using the same frequency that is located 4.6 kilometers away.

A higher or lower tolerance for interference may be tolerated by digital LMR systems. This either allows for an increased or decreased distance requirement for frequency reuse. An increase in the tolerance for interference would allow each transmitter site to install more channels as frequencies could be used more often in a limited geographic area. This results in a higher overall system capacity.

To increase the capacity of an ESMR system, smaller radio coverage areas are used in a geographic area. One of the simplest and least expensive ways to create smaller radio coverage areas is to divide the existing transmitter site's radio coverage areas into sectors. By using directional antennas, radio energy can be focused into specific areas (sectors) around the transmitter site. This can dramatically reduce the interference level to nearby transmitter sites that are located outside the directional radio beam focus. This reduced interference level allows the same frequencies to be used more often.

Figure 7.9 shows two transmitter sites that are divided (sectored) into six 60 degree sectors. Directional antennas are used to focus the radio energy so that only a portion of the site radio coverage area (e.g., 1/6th or 60 degrees) is used for each radio channel carrier. Although each sectored radio coverage area uses a different frequency, sectoring reduces interference with the other cells that are operating in the area. This allows nearby transmitter sites to reuse the same frequency.

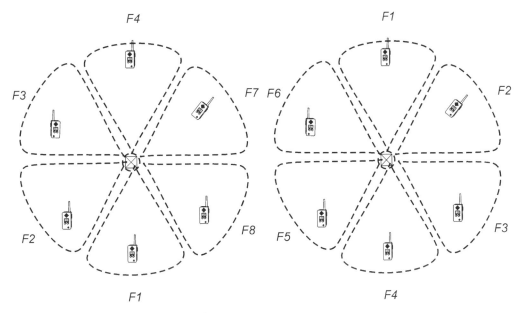

Figure 7.9, Radio Site Sectorization

Another method allows system expansion by the dividing of radio coverage sites into smaller radio coverage areas through the addition of transmitter sites. Radio coverage areas can be split by adding additional low power transmitter sites. Figure 7.10 shows the conversion of four existing radio coverage areas into six smaller radio coverage areas. To maintain low interference levels, the transmitter power levels at the original sites #2 and #3 are reduced and the two new transmitter sites, 2B and 3B, are added.

The maximum number of customers that can be serviced by a wireless system is primarily dependent on the number of installed radio channels and how often customers use (share) these channels. A radio system can serve a number of customers that is greater than its number of radio channels because each mobile radio usually accesses the system for only a few minutes each day. Dependent on the type of use (e.g., mobile data compared to wireless telephone use), approximately ten to forty mobile phone customers can be added to the system for each installed radio channel (typically 20 to 40 for mobile telephone use). If each radio site has 10 channels, it can provide service for 200 to 400 mobile telephone users.

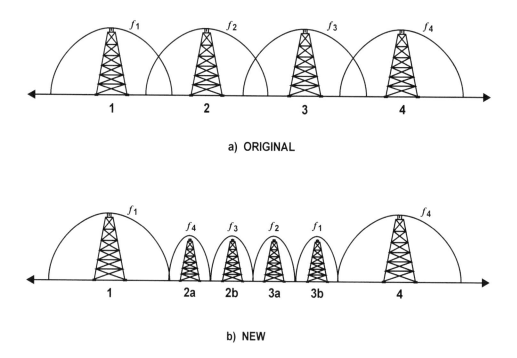

Figure 7.10, Radio Site Splitting

Handoff

As mobile radios move between radio coverage areas in advanced mobile radio systems, their calls may be automatically transferred between transmitter sites. This handoff process (sometimes called handover) allows a mobile radio to continuously communicate while moving through a large geographic area.

A typical handoff process is shown in Figure 7.11. First, the serving base determines that the mobile radio's signal strength falls below a minimum acceptable level (step 1). The base station then informs the system switching center or an adjacent base station that a handoff (call transfer) may be necessary. The adjacent base station (and other nearby base stations) begin to measure the mobile radio's signal strength. When an adjacent base station determines that the received signal strength is much higher than that of the serving base station (step 2), a command is sent to the mobile radio that instructs it to change to a frequency that is operating on the adjacent base station (step 3). Next (step 4), the mobile switching center transfers the voice communications path to the new base station to complete the call handoff. Early LMR systems did not offer handoff. Some systems use simulcast transmissions to cover large geographic areas, which eliminates the need for call transfer at the sacrifice of reduced system capacity.

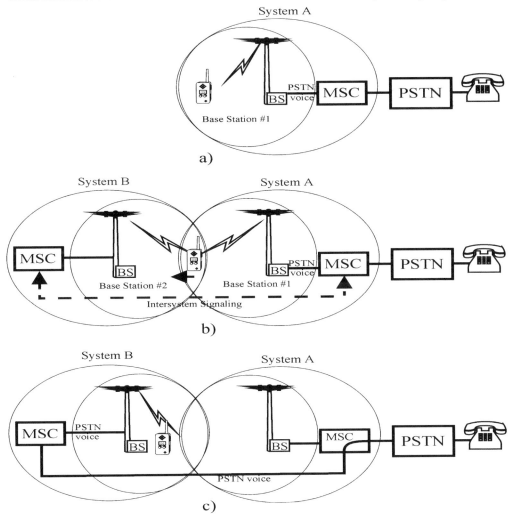

Figure 7.11, Radio Channel Handoff

Voting Receivers

To assist in the reception of the low power mobile radio signal, some wireless systems use more RF receivers than transmitters in a wide geographic area. These receivers (called voting receivers) are typically used because the system transmitter power can be much higher than the mobile radio's transmitter power. Because the system transmitters are at a higher power level, voice and message signals travel much further than can be received from the mobile radio. Through the use of multiple voting receivers, the reception from low power radios is enhanced by either selecting (voting) from the nearest receiver or by combining the signals from multiple receivers that are within range of the mobile radio's signal.

A system that uses voting receivers is shown in Figure 7.12. In this system, there is one transmitter that operates a base station with 100 Watts transmitter power. As the mobile radio moves throughout the transmitter site radio coverage area, the mobile radio continues to receive the signal from the same base station transmitter. However, because each mobile radio in the system can only transmit at 5 Watts, the system selects a receiver in the system that has the strongest received signal from the mobile radio. As the mobile radio moves from voting receiver #1 to voting receiver #2, the voting system will eventually select voting receiver #2 as the best choice to receive communication from the mobile radio.

100 Watts (base transmit)

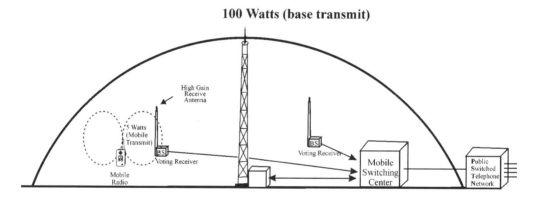

Figure 7.12, System with Voting Receivers

Upgrading Conventional LMR to Trunked Radio Systems

Trunked radio systems are much more efficient than conventional (two-way) radio systems. When conventional systems are converted to trunked radio systems, conventional mobile radios can either be replaced with trunking radios or conventional systems can share the trunked system.

Unlike traditional mobile telephone systems where there is exclusive use of radio channels within a specific geographic area, radio frequencies for LMR systems are often shared by multiple users. This complicates the process of conversion as a trunked radio system should detect or avoid causing interference to other users. When conventional systems share trunked systems, there needs to be a process of channel access control.

The sharing of radio channels in early LMR systems was often performed by the user waiting until the channel was free of activity (the user listening for no activity). Single radio channel frequencies were dedicated to groups of users such as taxi cab drivers. Because the users that are assigned to specific radio channels may have a need for significant usage while users on another channel would not, this limited the efficiency of the radio channel usage. Some radio channels in a specific geographic area could be unused while there may be several users waiting to share a single active channel in another area.

To help coordinate radio channel usage control, some conventional systems automatically indicate if a channel is busy through the use of squelch tones or codes. The users could be blocked from hearing or accessing the system until the lack of squelch tone or code indicated the channel was available.

When converting conventional radio channels to trunked radio channels, channel activity can be monitored to permit a mixing of conventional and trunked users in the same geographic area. When changing a system from conventional (two-way) to shared channel use, it is better to use one channel as a coordinating control channel while the others can be shared conventional and trunked channels. This control channel provides an access gateway for the trunked radio channel access. In this situation, existing users do not need to be part of the new trunked system. When activity is sensed on the trunked radio channel, the control channel automatically blocks the usage of that channel from other trunked users.

To convert conventional systems to trunked systems, a controller assembly needs to be added. In many cases, existing transmitters can be used and controllers can be installed that connect directly to the transmitter assemblies.

Integrated and Overlay

The conversion of LMR systems to support new technologies has two options: integration and overlay. Integrated systems use one network to control both existing (typically analog) and new technology (usually digital radio). As an alternative, two separate systems can independently exist that share radio channels. For overlay systems, the existing system may coordinate analog radio channel access and the overlay system could coordinate advanced digital radio channels. The use of an overlay system permits continued use of existing users and equipment.

Figure 7.13 shows that a conventional system can be converted to an integrated system through the addition or modification of a controller assembly. The controller assembly coordinates equipment control between older radio channel systems and new radio channel units. Integrated systems may allow multiple types of radio channel technology to be used in the same system (e.g., analog or conventional and digital or multiple types of digital). When using the integrated system technology and a new radio technology is to be added (e.g., DCMA or TDMA), the LMR system operator needs only to purchase additional radio cards with the new technology and upgrade the system software to control the new equipment. An overlay system requires that each system have its own control equipment. Although this diagram shows that system equipment for different technologies are co-located at a radio site, in some cases the transmitters of one system may share the common antenna. However, controller and radio transceivers are not usually compatible and may be located in different equipment racks.

Overlay systems are typical for LMR systems. When the overlay approach is used for LMR systems, the dispatch console or controller can be used to patch dissimilar technologies together.

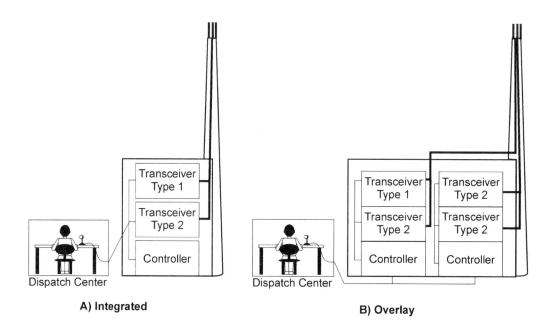

A) Integrated   B) Overlay

Figure 7.13, Integrated and Overlay Systems

Sub-Rate Multiplexing

To help increase the efficiency of network interconnections, advanced LMR systems sometimes multiplex multiple communication channels into a single communication line or circuit.
This is called sub-rate multiplexing.

Because digital radio systems reduce the data transfer rate through the use of speech data
compression, it is relatively easy for digital LMR systems to provide for sub-rate multiplexing.
Compressed voice signals range from 2 kbps to 16 kbps compared to the 64 kbps that is available for each landline communications channel.

The example in Figure 7.14 shows how eight compressed voice signals can share a single
DS0 (64 kbps) channel. In this system, an analog voice signal is converted to a PCM digital signal
at 64 kbps. A speech coder analyses and compresses the voice signal to 8 kbps. This compressed
digitized voice channel is time shared (time multiplexed) on a single 64 kbps DS0 channel with
eight other 8 kbps compressed voice channels. Each of the 64 kbps DS0 channels are then time
multiplexed onto a leased line (24 for T1 and 30 for E1 lines). The process is reversed on the
receiving end where the T1 or E1 channels are split back into DS0 channels. Then each DS0
channel is further demultiplexed again to produce eight 8 kbps compressed voice signals. These
digital signals are then converted back into their original 64 kbps signal by a speech decoder, and
a digital to analog converter then changes the digital signal back into its original analog form.

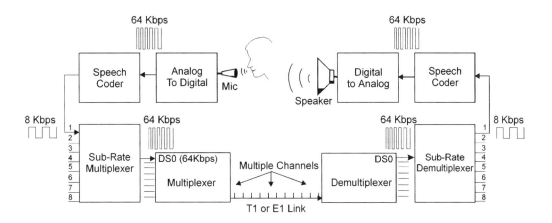

Figure 7.14, Sub-Rate Multiplexing

Distributed Switching

As systems become larger and transmitter sites become smaller, the switching requirements between radio sites increases exponentially. To help relieve the burden of rapid switching, some systems allow for localized switching in smaller geographic regions. This is called distributed switching.

The burden of switching on a main facility (maximum number of switching operations per second) is reduced by dividing the switching system into small geographic areas. For very large systems, switching between multiple radio sites and public telephone networks can be a very complex process.

Distributed switching also reduces the maximum capacity requirements of interconnection lines. For systems that have one centralized switch, all the remote sites must be connected back to the main switch. Distant radio sites are usually connected to the switching facility by cascading (hopping) the connection lines. This increases the maximum capacity requirements of interconnection lines that are located near the main switching facility.

Another advantage of distributed switching is the reduced effect of a system failure if a single switch in the network fails. When LMR systems are used for public safety (e.g., police and fire), system reliability is very important. The use of distributed switches can be structured to provide backup facilities in the event of a switch failure.

A sample ESMR network that uses distributed switching is shown in Figure 7.15. This example shows that some calls can be connected through the system without the need to use a main switching center. For example, two mobile radios are directly communicating with each other using communication path 3. Another example of distributed switching is the connection of a mobile radio to the PSTN without going through the main switch (path 2).

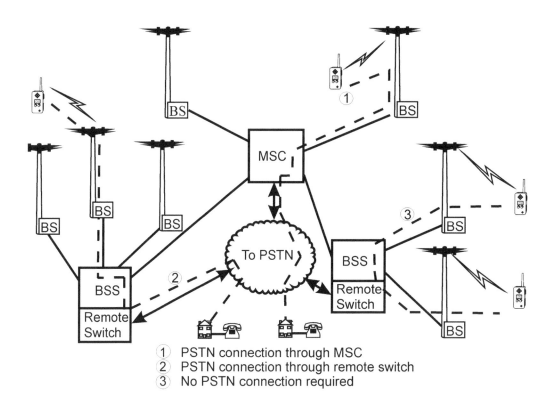

Figure 7.15, Distributed Switching

Echo Cancelers

Echoes that normally occur in analog systems are not as noticeable as echoes that are delayed in digital systems. This results in the use of advanced echo signal cancelers in digital LMR systems. Echoes are a more serious problem in digital systems because the amount of time it takes to process (compress) a digital audio signal adds delay to the transmitted signal. This increases the sensitivity to small echoes that occur with transmission (e.g., from a speakerphone or echo in a room).

Echo cancelers reduce the effect of delayed replicas of the desired signal through the use of advanced digital signal processing (DSP) techniques. Echo cancelers temporarily store small portions of an input signal and compare this signal to various delayed versions of the signal (intentionally delayed and stored). The delayed signal is analyzed to determine if part of the delayed signal has resulted from a delay of the original signal. Because the amount of delay and intensity (volume) of the delay is variable, determining if there is an echo signal is a very complex process. For example, a mobile radio may be using its speakerphone capability at high volume one minute (high volume, long delay) and it may shift to handset mode (low volume, low delay) the next minute. Because echo cancelers are only needed in certain interconnection situations, they are habitually grouped (pooled) in the switching center of the LMR system.

LMR System Interconnections

LMR systems may be connected to other types of networks. The common types of interconnections include the public telephone network, dispatch center, and data networks (such as the Internet). In addition to dedicated network system interconnections, other network components may be connected to an LMR system (such as a voice mail or messaging system).

Systems that are connected to a land mobile network are usually connected through a gateway. Gateways are devices or equipment that enable the adaptation of information between two dissimilar types of systems. A gateway may reformat data and/or insert new protocols that allow two dissimilar systems to communicate.

Public Telephone Network

One of the most common LMR interconnection options is the public switched telephone network (PSTN). Although connection to the PSTN is possible with almost any type of LMR system, government regulations may prohibit or limit the interconnection of an LMR system to the PSTN.

The public telephone network system is interconnected by different levels of switches. In the early development of the telephone system, switches were classified by their level of service. Lower level switches (called end office switches) were located closest to the end user (wired telephone) and higher level switches (called tandem switches) interconnected low level switches.

When ESMR networks are interconnected to the public telephone network, the connection is usually accomplished via a switching system. ESMR networks use a high speed digital switching system that can interconnect with the PSTN. These digital switches are capable of requesting call setup and termination through control signaling messages. These control messages are sent through a separate control signaling network that is used by other switches in the PSTN. This separate signaling system is called signaling system number seven (SS7). The SS7 signaling network is composed of databases (service control points (SCPs)) and packet data switching points (signaling transfer points (STPs)). SCPs are databases that are used by the telephone network to direct or instruct calls on how to be connected through the public telephone network. For example, an SCP may contain the destination address for a toll-free (800) number. An STP is used to route the control message packets through the network to their destination.

There are various types of standard connections between wireless networks and the public telephone network. The most common types that are used are Type 1 and Type 2. Type 1 is the most basic interconnection. Early LMR systems used this plain old telephone service (POTS) type of connection as it is the connection used by most standard telephones. This type of connection is also commonly used for dispatch consoles that do not require advanced public telephone network features (such as displaying the status of a telephone call). Type 2 connections have advanced control messaging capability that allows the exchange of telephone network control messages between the public telephone network and the land mobile radio network. Type 2 connections are also sub-categorized to provide various services. This includes connections dedicated for directory assistance, emergency service, and others. Figure 7.16 shows how land mobile networks are commonly connected to the public telephone network.

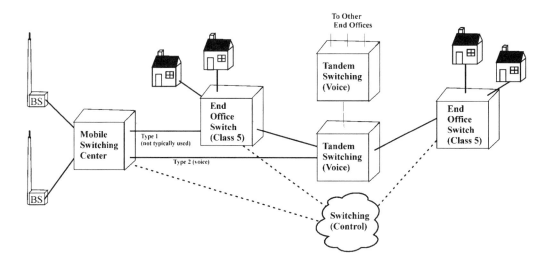

Figure 7.16, Public Telephone Network

In some countries, the public telephone network is divided into local exchange carriers (LECs) and long distance inter-exchange carriers (IXCs). LECs provide the local telephone service to end users and IXCs provide the long distance service between LEC companies. This tends to complicate the interconnection of LMR systems to public telephone networks as it may be possible for the wireless carrier to bypass the local telephone company.

Dispatch Center

Dispatch consoles are often used by conventional LMR networks to connect mobile radios to the public telephone network. When this is performed, it is called a "phone patch." To construct a phone patch, a dispatcher manually selects an open phone line and cross-connects it to the mobile radio audio line. Advanced LMR systems may allow automatic phone patching via radio signaling messages.

Dispatch centers contain both voice and control lines that can be interconnected with one or more radio site(s). A dispatch console may be located adjacent to the base station (such as in a taxi dispatch center) or at a remote location (such as at a public safety office). When a dispatch center is located at a remote facility, a radio link or leased telephone line may connect the dispatcher console to the radio site(s). Dispatch centers can also be used to cross-connect several different types of systems together. This may include public safety systems, phone lines, and internal communication system. The dispatch center operates like a manual switch system, allowing interconnection between audio lines.

Data Networks

One of the highest growth areas for LMR systems in the 2000s is the transfer between and interconnection to data networks. Interconnected mobile data networks include circuit switched data, packet switched data (such as the Internet) or short messaging services. Gateways adapt the communications between mobile radios and different types of data networks. Gateways may adapt data transfer to allow access to information databases such as picture files of wanted criminals, power utility grid maps, safety procedures and many others. Various companies specialize in creating the gateways to access these information databases.

Voice and Fax Mail Systems

Voice and fax mail systems temporarily store and forward audio or fax messages. A voice mail system appears to the network as an extension to a phone system. The voice mailbox automatically answers the call provided specific criteria is met (e.g., number of unanswered rings). The difference between a standard voice mail system and voice mail in an LMR system is the availability of additional mobile radio status information. For example, if a voice mail is waiting to be delivered, a voice mail waiting message may be sent to the mobile radio when the mobile radio turns on.

Customer Service Center

Customer care is required to answer technical and billing questions. Estimates show that in some systems, over 80% of all calls involve billing questions. LMR networks may also be set up to allow the automatic routing of calls to a customer service or billing center contingent on specific criteria (such as a customer's bill is not paid).

Customer service facilities can be part of the carrier's (service provider's) facilities or can overflow to another facility (outside service agency). ESMR systems can also use advanced intelligent networks (AIN) or computer telephony integration (CTI) to help route customer service requests. Through the use of CTI, it is possible to use the identification number of the mobile radio to automatically obtain billing records or customer care history.

System Design

Developing and growing an LMR system involves strategic network planning, radio propagation analysis, frequency planning and system testing.

Strategic Planning

One of the first steps in establishing an LMR system is the determination of the carriers key objectives and how to use these objectives to create implementation plans. These objectives ordinarily include how many mobile radios the system plans to serve (capacity), the quality level of service (coverage area and reliability) and cost objectives (operational and fixed costs).

The collection of demographic or user characteristics assists in the determination of system usage levels and locations of usage. This information allows the targeting of transmitter sites to provide service to key locations. The type of desired demographic information often includes the locations of main highways, industrial parks, convention centers, railway centers and airports. All of these offer possible high usage areas.

After an LMR technology has been selected (e.g., APCO project 25 or TETRA), equipment manufacturers and their systems are contacted and interviewed and purchase contracts are signed. As equipment installation progresses, validation testing is performed to ensure all planning objectives are achieved.

Radio Propagation

After the desired transmitter areas (radio coverage areas) are identified, using site locations and geographic maps of these areas, radio propagation analysis can be performed through the use of computer simulations. A typical radio site radio propagation simulation usually has different colors that indicate signal strength levels overlaid on maps of the area to be served. Although radio system simulations provide anticipated signal coverage and performance levels, temporary radio sites are often constructed at the anticipated location so actual testing can be performed.

Frequency Planning

Each transmitter is assigned one or more frequencies as part of a frequency plan. The frequencies and transmitter power levels are planned so that they do not cause interference with nearby transmitter sites that operate on the same frequency.

In addition to the requirement that radio transmitter frequencies are planned so as not to interfere with nearby radio sites, the assignment of radio channel frequencies must be separated by enough bandwidth that they do not interfere with each other in the same radio site. Radio channels in the same radio site can commonly be separated by as little as two to three channel bandwidths from other channels (commonly called the guard bandwidth). This requirement is a result of the small amount of radio energy that is transmitted outside the defined bandwidth of each radio channel. This interference is called adjacent channel interference. The minimum amount of channel separation is determined by a variety of factors including the type of modulation and power level.

System Testing and Verification

After a system begins to operate, it will require system testing and verification. System testing and verification determines RF coverage quality (e.g., to correct poor coverage areas) and correct system operation (e.g., to set the correct handoff levels). The signal quality is habitually tested for individual radio sites first. Next, adjacent radio sites are then measured to determine system performance. System operation is verified by measuring the signal quality level at handoff, blockage performance (number of calls that cannot get through) and the number of dropped calls.

To test and verify a radio system, a mobile test receiver is regularly used. The test equipment monitors and records the signal quality levels as a test vehicle moves throughout the system area. Both the signal quality level and position are recorded to allow the resultant data to be plotted later. These plots or graphs are used to determine where signal levels and handoff thresholds need to be increased and decreased.

The attenuation and signal fading characteristics are unique for each radio site, and they can change over time as a result of new construction or even the leaves falling off the trees. This results in a continuous effort to adjust radio transmission levels and frequencies.

Chapter 8
LMR Economics

Land mobile radio equipment costs have dropped by approximately 20% per year during the past seven to ten years. While the technology and mass production cost reductions for wireless telephones and systems are mature, new digital wireless telephones are more complex. However, when radios are produced in mass quantities, large sales volume promotes cost savings. System equipment costs for LMR equipment must compete against a mature, competitive analog and digital cellular equipment market, which already has the advantage of cost reductions due to large production runs. However, LMR systems are generally less expensive than cellular systems, ranging in price from several thousand to several million dollars.

The economic goal of a wireless network system is to effectively serve many customers at the lowest possible cost. The ability to serve customers is determined by the capacity of the wireless system. The key factors that determine the capacity of the system is the size of the radio sites related to one or more of the following: the ability to reuse frequencies, the spectral efficiency of the radio channels (the number of users that can share each radio channel) and the number of radio channels that can be installed in each radio site. For any radio access technology, system capacity is increased by the amount of spectrum that may be used and the number of radio units that can use each channel. If the number of radio sites remains constant, the efficiency of the radio access technology (e.g., the number of users that can share a single radio channel) and the ability to reuse frequencies (e.g., the number of radio channels in each radio site) determines the system capacity.

Wireless service providers usually strive to balance the system capacity with the needs of their customers. Running systems beyond their maximum capacity results in unavailable communication channels for their customers, while running systems that have excess capacity results in the purchase of system equipment that is not required, which increases cost. One of the key objectives of the new LMR technology is to achieve cost-effective service capacity, using techniques such as digital voice compression, very narrow radio channels or more efficient modulation.

Purchasing and maintaining wireless system equipment is only a small portion of the cost of a wireless system. Administration, leased facilities, and tariffs may play significant roles in the success of LMR systems.

The wireless marketplace is undergoing a change. New service providers, such as digital cellular technologies (e.g., GSM), are competing in the marketplace with new competitive features, such as group call. This is likely to increase wireless services competition. Sales and distribution channels may become clogged with a variety of wireless product offerings. Advanced wireless digital technologies offer a variety of new features that may increase the total potential market and help service providers to compete. These new features may offer added revenue and provide a way to convert customers to a more efficient digital service. The same digital radio channels that provide voice services may offer advanced messaging and telemetry applications.

Land Mobile Radios

According to the Strategis Group, the average wholesale cost of land mobile radios in 1998 was $880 for mobile mounted radios and $730 for portable units. This is much higher than the wholesale cost of commercial mobile radio (cellular and PCS) that ranged from $69 to $436 in the United States during 1998 [1]. The cost difference between private land mobile radios and commercial mobile radios is a result of economies of scale (high volume for cellular) and type of design (rugged for private land mobile radios). It is expected that the wholesale price of land

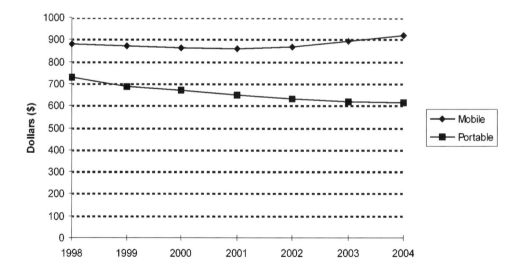

Figure 8.1, Land Mobile Radio Cost in the United States
Source: Strategis Group, Washington D.C.

mobile radios will drop to $613 by 2004 as systems become more standardized. Figure 8.1 shows the expected decline of the wholesale cost of land mobile radios in the United States.

Digital land mobile radios cost more than their equivalent analog counterparts due to the following primary factors: development cost, production cost, patent royalty cost, marketing, post-sales support and manufacturer profit.

Development Costs

Development costs are non-recurring costs that are required to research, design, test, and produce a new product. Unlike well-established FM technology, non-recurring engineering (NRE) development costs for digital LMR radios can be high due to the added complexity of digital design. Several companies have spent millions of dollars developing digital wireless products. Figure 8.2 shows how the non-recurring development cost per unit varies as the quantity of production varies from 20,000 to 100,000 units. Even small development costs become a significant challenge if the volume of production of digital wireless telephones is low (below 20,000 units). At this small production volume, NRE costs will be a high percentage of the wholesale price.

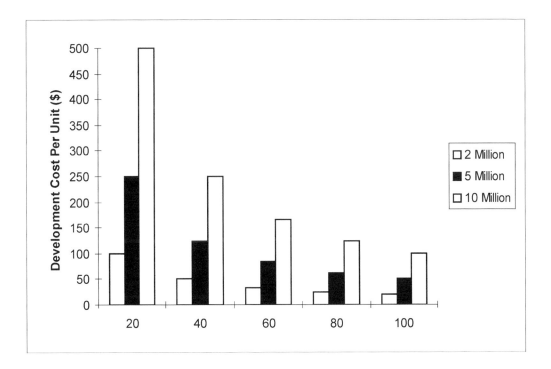

Figure 8.2, Mobile Radio Development Cost
Source: APDG Research

The introduction of a new technology presents many risks in terms of development costs. Some development costs that need to be considered include market research, technical trials and evaluations, industrial, electrical, and software design, prototyping, product and government type approval testing, creation of packaging, brochures, user and service manuals, marketing promotion, sales and customer service training, industry standards participation, unique test equipment development, plastics tooling, special production equipment fabrication and overall project coordination.

When a new product is created that is not a revision of an existing product using readily available components, cost-effective design is sometimes compromised. Cost-effective design is achieved by integrating multiple assemblies into a custom chip or hybrid assembly. Custom integrated circuit chip development is used to integrate many components into one low-cost part. Excluding the technology development effort, the custom Application Specific Integrated Circuit (ASIC) development usually requires a development setup cost that ranges from $250,000 to $500,000. There may be more than one ASIC used in a digital wireless telephone.

Cost of Production

The cost to manufacture a mobile radio includes the component parts (bill of materials), automated factory assembly equipment and human labor. Digital wireless radios are more complex than analog units. A digital mobile radio is composed of a radio transceiver and a digital signal processing section. The primary hardware assemblies that affect the component cost for a digital mobile radio are Digital Signal Processors (DSPs) and radio frequency assemblies. A single DSP, and several may be used, costs between $7 and $28 [2]. The Radio Frequency (RF) assemblies used in LMR radios require fast switching frequency synthesizers as compared to analog radios. In 1998, these RF components cost approximately $10-18. Other components that are included in the production of a mobile radio include printed circuit boards, integrated circuits and electronic components, radio frequency filters, connectors, a plastic case, a display assembly, a keypad, a speaker and microphone and an antenna assembly. In 1998, the bill of materials (parts) for a digital mobile radio was approximately $140 [3].

The assembly of wireless radios requires a factory with automated assembly equipment. Each production line can cost between two and five million dollars. Regularly, one production line can produce a maximum of 500-2,000 units per day (150,000-600,000 units per year). The number of units that can be produced per day depend on the speed of the automated component insertion machines and the number of components to be inserted. Normally, production lines are often shut down one day per week for routine maintenance and two weeks per year for major maintenance overhauls which leaves about 300 days per year for the manufacturing line to produce products. Between interest cost (10-15% per year) and depreciation (10-15% per year), the cost to own such equipment is approximately 25% per year. This results in a production facility overhead of $500,000 to $1.25 million per year for each production line. Figure 8.3 shows how the cost per unit drops dramatically from approximately $10-25 per unit to $1-3 per unit as volume increases from 50,000 units per year to 400,000 units per year.

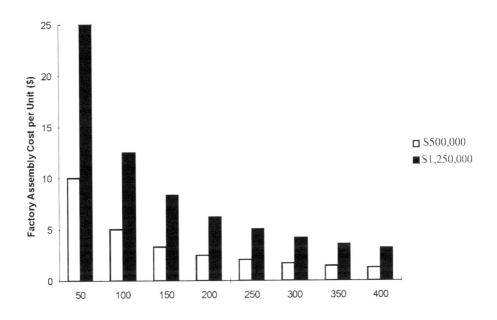

Figure 8.3, Mobile Radio Factory Assembly Equipment Cost
Source: APDG Research

While automated assembly is used in factories for the production of mobile radios, there are some processes that require human assembly, such as the placement of large bulky components (e.g., radio filters). Efficient assembly of a mobile radio in a modern factory requires 1/2 to 1 hour of human labor. The amount of human labor is a combination of all workers involved with the plant, including administrative workers and plant managers. The average loaded cost of labor (wages, vacation, insurance) varies from approximately $20-$40 per hour, based on the location of the factory and the average skill set of human labor. This results in a labor cost per unit that varies from $10-$40. Since digital mobile radios may have more parts to assemble due to the added complexity, the labor cost may increase.

Patent Royalty Cost

Another significant cost factor to be considered is patent royalties. Several companies have disclosed that they believe they have some proprietary technology that is required to implement some features defined in some of the LMR standard specifications. Many large manufacturing companies exchange the right to use their patented technology with other companies that have patented technology they want to use. Patents from other companies that may be desirable or essential to implement the standard specifications may not have been discovered or disclosed.

Marketing Cost

The marketing cost, which is included in the wholesale cost of the mobile radios, includes a direct sales staff, manufacturer's representatives, advertising, trade shows and industry seminars. Worldwide, operators report that more customers are recruited from their own sales force than by any other means.

Mobile radio manufacturers ordinarily dedicate a highly paid representative or agent for key customers. Much like the sales of other consumer electronics products, manufacturers employ several technical sales people to answer a variety of technical questions prior to the sale.

There are mobile radio manufacturers that use independent distributors to sell their products. This practice is more prevalent for smaller, lesser-known manufacturers who cannot afford to maintain dedicated direct sales staff. These representatives commonly receive up to 4% of the sales volume for their services.

Advertising programs used by wireless radio manufacturers involve broad promotion for brand-recognition and advertisements targeted for specific products. The typical advertising budget for mobile radio manufacturers varies from approximately 3%-6%. The budget for brand recognition advertising normally ranges from less than 1% to over 4%. Product-specific advertising is often performed through cooperative advertising, which is paid from the manufacturer to a distributor or retailer. The amount paid commonly varies from 2%-4% of the cost of the products sold to the distributor. For the distributor or retailer to receive the cooperative commission, they must meet the manufacturer's advertising requirements. This approach allows distributors and retailers to determine the best type of advertising for their specific markets.

Mobile radio equipment manufacturers ordinarily exhibit at trade shows three to four times per year. Trade show costs are high. Mobile radio manufacturers exhibiting at trade shows normally have large trade show booths, gifts, and theme entertainment. Hospitality parties at trade shows are also common. Mobile radio manufacturers often bring 5-20 sales and engineering experts to the trade shows to answer distributor questions.

To help promote the industry and gain publicity, wireless radio manufacturers participate in a variety of industry seminars and associations. The manufacturers regularly have a few select employees who write for magazines and speak at industry seminars. All of these costs, and others, result in a combined estimated marketing cost for mobile telephone manufacturers of 10%-15% of the wholesale selling price.

Post-Sales Support

The sale of mobile radios involves a variety of costs and services after the sale of the product (post-sales support), including warranty servicing, customer service and training. A customer service department is required for handling distributor and customer questions. Because the average customer for a mobile radio is not technically trained in radio technology, the amount of non-technical questions can be significant. Distributors and retailers require training for product feature operation and servicing. The post-sales support cost for wireless radios is usually between 4%-6%.

Manufacturer's Profit

Manufacturers must make a profit as an incentive for manufacturing products. The amount of profit a manufacturer can make, as a rule, depends on the risk involved with the manufacturing of products. As a general rule, the higher the risk, the higher the profit margin.

The mobile radio market in the early 1990s became very competitive due to manufacturers' abilities to reduce cost through mass production and through competition with digital cellular telephone service. To effectively compete, manufacturers had to invest in factories and technology, which increased the risk and the required profit margin. In 1998, the estimated gross profit in the wireless radio manufacturing industry was approximately 10%-30%.

System Equipment

The cost for wireless system equipment includes the following primary factors: development cost, production cost, patent royalty cost, marketing, post sales support and manufacturer profit.

Development Costs

The mobile radio network system equipment development costs are much higher than wireless telephone development costs. When a completely new technology is introduced, wireless network system development costs can exceed millions of dollars because the complexity of an entire wireless system is significantly greater than a mobile telephone. Thus, more testing and validation is required. This high investment would limit most manufacturers from producing products for the new technology.

While base station radios perform similarly to wireless telephones, the coordination of all the base stations involves many additional electronic subsystems. Additional assemblies include communication controllers in the base station and switching center, scanning locating receivers or voting receivers, communication adapters, switching assemblies and large databases to hold customer features and billing information. All these assemblies require hardware and very complex software.

LMR systems have a unique feature that has enticed many manufacturers to develop network products. For earlier LMR systems, the equipment in the network was usually unique to the manufacturer. This required a manufacturer to develop an entire network. This would include the switching system, software, base station radios and controllers. The LMR specification has defined in detail many of the network parts. Most network parts in a LMR system are not unique to a single manufacturer. This allows a manufacturer to only develop parts of the network. Thus, the initial investment is lower, and more manufacturers have produced system equipment. This has increased the number of suppliers and decreased the cost of system equipment.

Unlike mobile telephones, when a mobile radio system develops a problem, the entire system can be affected. New hardware and features require extensive testing. Testing LMR systems can require thousands of hours of labor by highly skilled professionals. Introducing a new technology is much more complex than simply adding a new feature.

Cost of Production

The physical hardware cost for digital LMR network system equipment should be more expensive than analog network system equipment due to the added technological complexity. However, the physical hardware cost for digital LMR network system equipment may actually be less than older analog equipment due to more standardized products and economies of scale.

The cost to manufacture an LMR network system includes the component parts, automated factory equipment and human labor. Because the number of mobile radios that share a single LMR radio channel can exceed 200 (not all users access the system at the same time), the quantity of wireless network system assemblies produced is much smaller than the number of wireless radios. Setting up automated factory equipment is time consuming. For small production runs, much more human labor is used in the production of assemblies because setting up the automated assembly is not practical. The production of system equipment involves a factory with automated assembly equipment for specific assemblies. However, because the number of units produced for system equipment is normally much smaller than mobile radios, production lines used for LMR system equipment are often shared for the production of different assemblies or remain idle for periods of time.

With over 60 countries using LMR systems, the demand for LMR system equipment is increasing exponentially. This increased demand allows for larger production runs, which reduce the average cost per unit. Large production runs also permit investment in cost-effective designs, such as using Application Specific Integrated Circuits (ASICs) to replace several individual components.

The maturity of digital radio technology is promoting cost reductions through the use of cost-effective equipment design and low-cost commercially available electronic components. In the early 1990s, many technical system equipment changes were required due to changes in radio specifications. Manufacturers had to modify their equipment based on field test results. For example, complex echo cancelers were required due to the long delay time associated with digital speech compression. Manufacturers ordinarily did not invest in cost-effective custom designs because of the rapid changes. As the technology has matured, the investment in custom designs is possible with less risk. In the early 1990s, it was also unclear which digital technologies would become commercially viable, limiting the availability of standard components. Today, the success of digital systems has created a market of low-cost digital signal processors and RF components for digital LMR systems.

Like the assembly of mobile radios, the assembly of system radio and switching equipment involves a factory with automated assembly equipment. The primary difference is smaller production runs, multiple assemblies, and more complex assembly.

The number of equipment units that are produced is much smaller than the number of mobile radios produced because an LMR radio channel can serve 150 to 200 customers. The result is much smaller production runs for LMR network system equipment. While a single production line can produce a maximum of 500-2,000 assemblies per day [4], several different assemblies for radio base stations are required. A change in the production line from one assembly process to another can take several hours or several days. LMR system radio equipment requires a variety of different connectors, bulky RF radio parts and large equipment case assemblies. Due to lower production volumes and many unique parts, it is not usually cost-effective to use automatic assembly equipment. For unique parts, there are no standard automatic assembly units available. Because of this more complex assembly and the inability to automate many assembly steps, the amount of human labor is much higher than for mobile radios.

Each automated production line can cost two to five million dollars. The number of units that can be produced per day varies depending on the speed of automated component insertion machines, the number of components to be inserted, the number of different electronic assemblies per equipment, and the amount of time it takes to change/setup the production line for different assemblies. If it is assumed there are four electronic assemblies per base station radio equipment (e.g., controller, RF section, baseband/diagnostic processing section and power supply), the automated production cost for base station equipment should be over four times that of mobile radios.

Figure 8.4 shows how the production cost per unit drops dramatically from approximately $400-$1,000 per unit to $50-$125 per unit as the volume of production increases from 5,000 units per year to 40,000 units per year. This chart assumes production cost is four times that of mobile radios due to the added complexity and the use of multiple assemblies.

While automated assembly is used in factories for the production of mobile radios, there are some processes that require human assembly. Efficient assembly of base station units in a modern factory requires between five and ten hours of human labor. The amount of human labor includes all types of workers from administrative workers to plant managers. The average loaded cost of labor (wages, vacation, insurance) varies from approximately $20-40 per hour, which is based on the location of the factory and average workers skill set. The resultant labor cost per unit varies from $100-400.

Patent Royalty Cost

There are fewer manufacturers that produce wireless network system equipment because the use of many different technologies is involved. Large manufacturers have a portfolio of patents that are commonly traded. Cross licensing is common and tends to reduce the cost of patent rights. When patent licensing is required, the patent costs are sometimes based on the wholesale price of the assemblies in which the licensed technology is used.

Marketing Costs

Marketing costs that are included in the wholesale cost of wireless system equipment include a direct sales staff, sales engineers, advertising, trade shows, and industry seminars. Wireless system manufacturers often dedicate several highly paid representatives to key customers. Wireless system sales are much more technical than the sale of mobile radios. Manufacturers employ several people to answer a variety of technical questions prior to the sale.

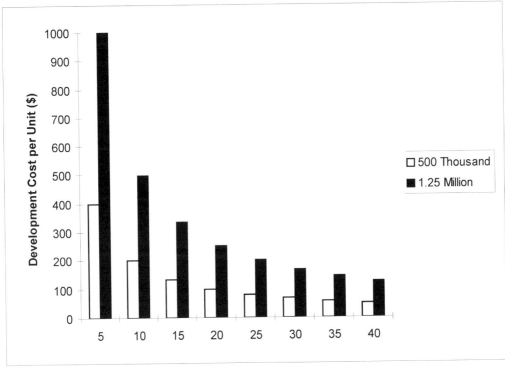

Figure 8.4, System Equipment Factory Assembly Equipment Cost
Source: APDG Research

Advertising used by LMR system equipment manufacturers involves broad promotion for brand recognition and advertisements targeted for specific products. The budget for brand recognition advertising is commonly small and is targeted to specific communication channels because the sale of wireless system equipment involves only a small group of people who usually work for an LMR service provider. Product-specific advertising is also limited to industry-specific trade journals. Much of the advertising promotion of wireless system equipment occurs at trade shows, industry associations, and direct client presentations. The advertising budget for wireless system equipment manufacturers is regularly less than 2%.

LMR system manufacturers also exhibit at trade shows normally three to four times per year. Wireless system equipment manufacturers exhibiting at trade shows often have large hospitality parties that sometimes entertain hundreds of people. Wireless system manufacturers often bring 15-40 sales and engineering experts to the trade shows to answer customer questions.

To help promote the industry and gain publicity, LMR system manufacturers participate in many industry seminars and associations. These manufacturers use trained experts to present at industry seminars. All of these costs result in an estimated marketing cost for system equipment manufacturers of approximately 8-10% of the wholesale selling price.

Post-Sales Support

The sale of LMR systems involves a variety of costs and services after the sale of the product. This includes warranty servicing, customer service, and training. A seven day, 24-hour customer service department is required for handling customer questions. Customers require a significant amount of training for product operation and maintenance after a system is sold and installed. The post-sales support costs for wireless system equipment is ordinarily 3-5%.

Manufacturer's Profit

Standardization of systems and components, particularly digital LMR, has led to a rapid drop in the wholesale price of system equipment. While the increased product volume of LMR system equipment has resulted in decreased manufacturing costs, the gross profit margin for wireless system equipment has decreased. The estimated gross profit in the mobile radio system equipment manufacturing industry is 10-15%.

System Capital Cost

The LMR service provider's investment in network equipment includes radio sites, base station radio equipment, dispatch or switching centers, and network databases. One primary objective of the new digital LMR technologies is to decrease the network cost per customer (capital cost per customer), which is possible because the new technologies can serve more customers with less physical equipment.

In theory, existing analog LMR technology can serve an almost unlimited number of customers in a designated area by replacing large radio site areas with many Microcells (small cell coverage areas). However, expanding current analog systems in this way increases the average capital cost per customer, due to the added cost of increasing the number of small radio sites and interconnection lines to replace a single large radio site. For example, when a radio site with a 15-km radius is replaced by radio sites with a ½-km radius, it will take over 700 small radio sites to cover the same area.

One of the reasons that digital LMR technologies were developed was to allow for cost-effective capacity expansion. Cost-effective capacity expansion results when existing radio sites can offer more communication channels, which allows more customers to be served by the same radio site. As systems based on such new technologies expand, the average cost per customer decreases.

Radio Site

The radio site is composed of a radio tower, antennas, a building, radio channels and system controllers. The radio tower is commonly 100-300 feet tall. The cost ranges from $30,000-$300,000. While some of the largest towers can cost $300,000, an average cost of $80,000 is typical because, as systems expand, smaller radio towers can be used.

A building on the radio site property is required to store radio site equipment. This building must be bullet proof, have climate control and have various other non-standard options. The estimated building cost is $40,000.

Many radio sites can be located on a very small area of land. Land is either purchased or leased. In some cases, existing tower space can be leased for $500-$1,000 per month. If the land is purchased, the estimated cost of the land is approximately $60,000.

Radio sites are not usually located where high-speed telephone communication lines are available. Typically, it is necessary to install a T1 or E1 trunk communications line to the radio site that is leased from the local phone company. If a microwave link is used in place of a leased communication line, the communications line installation cost will be applied to the installation of the microwave antenna. The estimated cost of installing a T1 or E1 communications line is approximately $10,000.

The land where the radio site is to be located must be cleared, foundations must be poured and the fencing, building and tower must be installed. A construction cost of $50,000 is estimated (plus there is the added expense of governmental regulations which will vary by country). In addition, a backup power supply (generator) may be included ($10,000 estimated). Figure 8.5 shows the estimated cost for a typical radio site without the radio equipment.

In addition to the tower and building cost, radio equipment must be purchased. Each LMR radio site usually has several radio channels. To determine the total number of customers that can be serviced by a radio site, the number of radio channels are multiplied by the number of customers that can be serviced per radio channel. Because not everyone uses every radio channel at the same time, LMR systems that are used for PSTN interconnection service regularly add 20-40 (30 average) customers per voice channel. This is called the loading of the system. LMR systems that are used for dispatch service may add hundreds of users per radio channel.

Item	Cost (in thousands)
Radio Tower	$80
Building	$40
Land	$60
Install Communication Line	$10
Construction	$50
Backup Power Supply	$10
Total	$250

Figure 8.5, Estimated Radio Site Capital Cost Without Radio Equipment
Source: APDG Research

A digital LMR radio channel may allow three or more customers to simultaneously use each radio channel. With an average of three voice channels per radio channel (by time division or code division multiplexing) and an average loading of 30 customers per voice channel, each radio channel allows the LMR service provider to add 90 additional customers. Digital systems that allow up to six users to share each radio channel permits LMR service providers to add 180 additional customers per radio channel.

If there are an average of five radio channels installed in a single LMR radio site, each analog radio site allows approximately 150 customers. Digital systems that allow three users to share each radio channel allow 450 customers to be added to the system. Digital systems that allow four users to simultaneously share a channel allow up to 600 customers to be added for each radio channel. If the system allows six users to share each channel, the operator can add 900 customers. This means a LMR system that has 20 radio sites with an average of five radio channels per radio site can serve approximately 3,000 customers for analog, 9,000 customers for 3x digital, and 18,000 customers for 6x digital.

After the total investment of each radio site is determined, the radio site capital cost per customer can be determined by dividing the total radio site cost by the number of customers that will share the resource (radio site).

Digital radio channels are more expensive. More efficient radio channels require advanced signal processing. APDG research estimates the cost for radio channels that can provide (multiplex) three simultaneous users per radio channel to be $15,000, four users per radio channel to be $18,000 and six users per radio channel to be $20,000.

Figure 8.6 shows the average capital cost per customer. In column one, we see analog technology that supports one voice channel per carrier. If the average cost of a radio channel for analog LMR technology is $10,000 (this includes the line adapters and a proportion of the controller cost) and five radio channels are installed per radio site, the total radio equipment cost is approximately $50,000. The tower and building costs are added to this, bringing the total radio site cost to approximately $300,000. Column two shows that the cost of a 3:1 digital radio is approximately $15,000, and the total cost of the radio equipment and tower site is approximately $325,000. For the 4:1 system (column 3), the radio channel cost increases, bringing the combined radio equipment and tower site cost to $340,000. For the 6:1 system (column 4), the radio channel cost increases to approximately $20,000, resulting in a total transmitter site cost near $350,000.

If the average number of customers that can be added for each analog radio site is 150, the average capital cost for a new customer is $2,000. Because the average number of customers that can be served by a single radio site increases with digital technology, the average capital cost for a new customer for digital systems is $722 for 3:1 digital, $567 for 4:1 digital and $389 for 6:1 digital.

The ability of radio channels to serve more customers through one piece of RF equipment reduces the number of required RF equipment assemblies, power consumption and system cooling requirements. Sharing a channel in this way (multiplexing) also reduces the radio site size and backup power supply (generator and battery) requirements, and ultimately, provides additional cost savings.

Category	Analog	3:1 Digital	4:1 Digital	6:1 Digital
Cost per RF Radio Channel	$10,000	$15,000	$18,000	$20,000
Number of Radio Channels per Cell Site (3 sector)	5	5	5	5
Total Radio Channel Cost	$50,000	$75,000	$90,000	$100,000
Tower and Building Cost	$250,000	$250,000	$250,000	$250,000
Total Cell Site Cost	$300,000	$325,000	$340,000	$350,000
Number of Voice Paths per Radio Channel	1	3	3	6
Number of Subscribers per Voice Channel	30	30	30	30
Number of Subscribers per Cell Site	150	450	450	900
Cell Site Capital Cost per Subscriber	$2,000	$722	$567	$389

Figure 8.6, Radio Site Capital Cost per Customer
Source: APDG Research

System Operations Center

Radio sites are usually connected to a dispatch center or intelligent switching system (called the "switch"). An estimated capital cost of $100 per customer is used for the dispatch center or switch equipment and its accessories.

The dispatch or switching center must be located in a long-term location (10-20 years), usually near a Public Switched Telephone Network (PSTN) central office switch connection. The building contains switching and communication equipment. Commonly, customer databases are located in the switching center facility. The switching center software and associated LMR system equipment commonly contain basic software that allows normal mobile radio operation (place and receive calls). Special software upgrades that allow advanced services are available at additional cost.

System Operational Cost

The costs of operating an LMR system include leasing and maintaining communication lines, local and long distance tariffs, billing, administration (staffing), maintenance and fraud. The operational cost benefits of installing digital equipment includes a reduction in the total number of leased communication lines, a reduction in the number of radio sites, a reduction in maintenance costs, and a reduction of fraud due to advanced authentication procedures.

Leasing and Maintaining Communication Lines

A system operator must lease communication lines to run between radio towers or must connect radio sites by installing and maintaining microwave links between them. The typical cost for leasing a 24-channel line between radio sites in the United States in 1998 ranged from $300-$550 per month ($500 typical) [5]. The average cost of leasing a 30-channel E1 line in Europe ranged from $450 to over $2,000 per month ($800 typical) [5]. Leasing cost for communication lines depends on the distance of connection points and guaranteed grade of service. The greater the distance, the higher the cost. Because of increased competition in the telecommunications area by cable companies and competitive local exchange carriers offering high speed digital leased lines, it is anticipated that the average cost of leased E1 and T1 lines will continue to decrease.

Installing microwave radio equipment can eliminate the monthly cost of leased lines. The cost of microwave radio links between two radio sites ranges from approximately $20,000-$100,000.

Similar to the sharing on a radio channel, the number of customers that can share the cost of a communication line (loading of the line) varies with the type of service. For cellular-like customers who ordinarily use the phone for two minutes per day, approximately 600 customers can share a T1 (25 customers per voice path x 24 voice paths per communication line) or 750 customers per E1 (25 customers per voice path x 30 voice paths per communication line). This can be compared to residential-type service where customers use the phone for approximately 30 minutes per day. For residential service, approximately 120 customers can be loaded onto a T1 or 150 per E1. Average office customers use the phone for approximately 60 minutes per day. For office usage, approximately 60 customers can be loaded onto a T1 or 75 onto an E1.

The monthly cost per customer is determined by dividing the monthly cost by the total number of customers. Figure 8.7 shows the estimated monthly cost for interconnection charges for T1 and E1 communication lines. The estimated monthly cost is based on 100% use of the communication lines. If the communication lines are not used fully (it is rare that communication lines are used at full capacity), the average cost per line increases.

Digital signal processing for all the proposed technologies allow for a reduction in the number of required communications links through the use of sub-rate multiplexing. Sub-rate multiplexing allows several users to share each 64,000 bit per second (kb/s) communications (DS0/PCM) channel because digital voice information is compressed into a form much smaller than the existing communication channels. If 8 kb/s speech information is sub-rate multiplexed, up to 8 voice channels can be shared on a single 64 kb/s channel, which can reduce the cost of leased lines significantly.

Service Type	Type of Leased Line	T1 Line Cost per Month	Number of Channels per Line	Customers per Channel	Customers per Line	Total Cost per Month
Cellular	T1	$500	24	25	600	$0.83
LEC (residential)	T1	$500	24	5	120	$4.17
Office	T1	$500	24	2.5	60	$8.33
Cellular	E1	$800	30	25	750	$1.07
LEC (residential)	E1	$800	30	5	150	$5.33
Office	E1	$800	30	2.5	75	$10.66

Figure 8.7, T1 and E1 Monthly Communications Line Cost
Source: APDG Research

Local and Long-Distance Tariffs

Telephone calls in LMR systems are often connected to other local and long-distance tele-phone networks. When LMR systems are routed to existing landline telephone customers, they are normally connected through the local wired telephone network. The local telephone compa-ny regularly charges a small monthly fee and several cents per minute (approximately 3 cents per minute) for each line connected to the LMR carrier. Because each mobile radio customer uses the mobile radio for only a few minutes per day to connect to outside lines, the LMR service provider can use a single connection (telephone line) to the PSTN to service hundreds of cus-tomers.

In the United States and other countries that have separate long-distance service providers, when long-distance service is provided through a local telephone company (LEC), a tariff is paid from its LMR service provider to the local exchange company (LEC). These tariffs can be up to 45% of the per minute charges for long-distance service. Due to government regulations limiting the bundling of local and long-distance service, it is necessary for some LMR service providers to separate their local and long-distance service. Recent regulations may permit LMR carriers to bypass the LEC and save these tariffs.

Billing Services

Billing involves gathering and distributing billing information, organizing the information and invoicing the customer. As customers initiate calls or use services, records are created. Each billing record contains details of each billable call, including who initiated the call, where the call was initiated, the time and length of the call and how the call was terminated. Each call record contains approximately 100-200 bytes of information [6]. The billing records are usually stored in the company's own database. If the mobile radio customer is allowed to visit another system, billing records must be exchanged. Until the the early 1990s, LMR systems were not interconnected. This required the use of a magnetic tape to transfer records to standard Automatic Message Accounting (AMA) format. Today, billing records are regularly sent directly to a clearinghouse company to accumulate and balance charges between different LMR service providers.

With the introduction of advanced services, billing issues continue to become more complicated. The service cost may vary between different systems. To overcome this difficulty, some service providers have agreed to bill customers at the billing rate established in their home system.

Each month, billing records must be totaled and printed for customer invoicing, invoices must be mailed and checks must be received and posted. The cost for billing services is approximately $1 per month. Billing cost includes routing and summarizing billing information, printing the bill and mailing the bill. To help offset the cost of billing, some LMR service providers have started to bundle advertising literature from other companies along with the invoice. To expedite the collection, some LMR service providers offer direct billing to bank accounts or charge cards.

Operations, Administration and Maintenance (OA&M)

Running an LMR service company requires people with many different skill sets. Staffing requirements include executives, managers, engineers, sales, customer service, technicians, marketing, legal, finance, administrative, and other personnel to support vital business functions. The present staffing levels for LMR operators range from 10 to 25 employees for each 10,000 customers. If we assume a loaded cost (salary, expenses, benefits and facility costs) of $40,000 per employee, this results in a cost of $3.33 to $8.33 per month per customer ($40,000/10-25 employees/10,000 customers/12 months).

Maintenance and repair of wireless systems is critical to the revenue of a wireless system. In large systems, staffs of qualified technicians are hired to perform routine testing. Smaller LMR systems often have an agreement with their manufacturer or another wireless service provider to provide these technicians when needed. Most new LMR systems have automatic diagnostic capabilities to detect when a piece of equipment fails. These wireless systems usually have an automatic backup system, which can provide service until the defective assembly is replaced.

Land and Site Leasing

In rural areas, exact locations for radio site towers are not required. The result is that land leasing is not a significant problem. In urban areas and as systems mature, more exact locations for radio sites are required. This results in increased land leasing costs. By using a more efficient RF technology, one radio site can be used to serve more channels, which limits the total number of required radio sites.

Land leasing is, as a rule, a long-term lease for very small portions of land (40-200 square meters) for approximately 20 years or longer. The cost of leasing land is dependent on location. Premium site locations such as sites on key buildings or in tunnels can exceed the gross revenue potential of the radio site.

Another leasing option involves leasing space on an existing radio tower. Site leasing on an existing tower is approximately $500 per month. Site leasing eliminates the requirement of building and maintaining a radio tower.

Service Revenue Potential

In mid 1998, there were over 16 million LMR customers in the United States [source: Strategis Group]. While in the United States the average LMR bill is not much higher than the average wired residential telephone bill (approximately $50 per month), the amount of usage for an LMR radio is approximately one-tenth of residential usage. Outside the United States, flat rates range from $10-72.50 [Source: IMTA].

The LMR monthly bill in the U.S. has declined each year over the last five years. The average charge per minute has not decreased much; however, the amount of usage has decreased because new customers entering the market are consumers who do not use their mobile radios very much. Outside the United States, operators do not typically charge per minute of use. However, when they do, they often charge different rates for interconnect service and dispatch service. In many countries, the average charge per minute has declined due to competition from other wireless services.

The rapid growth of LMR markets is due to increased awareness of LMR service, new system availability and new service offerings in more than 55 countries. Because many new LMR systems are starting, it is reasonable to assume a continued yearly growth of over 45% per year for most LMR markets. In addition to the sale of LMR radios to normal-use consumers, it is possible to sell many times as many handsets for other uses, such as utility meter reading, vehicle tracking, environmental sensing and other applications, which have the potential for sales in excess of millions each worldwide.

The main revenue for LMR service providers is derived from providing telecommunications service. In 1998, a majority of the service revenue came from voice services. Digital wireless systems provide for increased service revenue that comes from a variety of sources, such as advanced services and system cost reduction.

System Cost to the Service Provider

One of the advantages of digital service is that it allows more customers to share the same system equipment. Financing LMR system equipment (network equipment cost) accounts for approximately 10-15% of the service provider's cost. Digital wireless systems can offer a reduction of approximately 60% of system equipment cost per customer. Some of the advanced features of digital wireless also provide for reductions in Operations, Administration and Maintenance (OA&M) costs. These system cost reductions offered by digital wireless technology allow LMR service providers to more effectively compete against other established wireless service providers, such as cellular and PCS.

Voice Service Cost to the Consumer

Over the past few years, the average cost of airtime usage to an LMR customer has not changed very much. To help attract customers to digital service, some LMR service providers have offered discounted airtime plans to high usage customers. This discount provides a significant incentive to high usage customers.

Data Service Cost to the Consumer

There are two types of data services that are available to customers: continuous (called "circuit switched data") or brief packets (called "packet switched data"). Commonly, continuous data transmission airtime usage is charged at the same rate as voice transmission. Packet data transmission is often charged by the packet or by the total amount of data that has been transferred. When charging by the packet (approximately 30 bytes), the cost per packet ranges from approximatel 1/5 cent to over 1/2 cent per packet. When charging by the kilobyte of information, the cost ranges from approximately 7 cents to over a dollar per kilobyte.

Mobile Radio Cost to the Consumer

In 1984-85, mobile radio prices in the United States varied from $2000-$2500. By 1997, a customer could choose from a variety of mobile radios for $500 to more than $2,000 [Radio Resource 1999 Specs Survey]. Outside the United States, the cost of radio units has been decreasing due to competition from other wireless services and other trunking operators. One common barrier to using LMR service typically has been its high radio unit price and the lack of manufacturer subsidies, which are available for cellular telephones but not for radios. It is not typical to be able to purchase a mobile radio for one cent when a company signs up for service. Depending on the type of radio, research shows that mobile radios tend to be less expensive in Latin American countries, more expensive in European countries, and most expensive in Asia. Over the years, terminal prices have decreased due to factors specific in each country. This trend can only help to increase overall subscribers. The primary reasons for the continued high growth of the mobile radio market are low usage fees, declining terminal equipment costs and stable or reduced usage charges.

New Features

Customers purchase mobile radios based on their own value system, which estimates the benefits they will receive. Although a strong feature is the low usage cost provided by LMR systems, new features provide new benefits to the consumer. These features can be used for product differentiation and to increase service revenue. Key new features available for LMR systems include digital voice quality, voice privacy, messaging services, longer battery life and many others. These new features are used to persuade customers to convert from using an analog radio or to pay extra for these new advanced services.

While the first analog LMR radio weighed over 80 pounds and required almost all available trunk space, mobile radio technology has evolved to allow handheld mobile radios. The first digital mobile radios were only slightly larger than their analog counterparts. The size of digital mobile radios continues to be reduced as production volumes allow for custom integrated circuit development, which integrates the analog and digital processing sections. Digital mobile radios are approaching and will surpass the race over analog for ever-smaller radios [8].

New features allow for different types of customers. With advanced data capabilities, LMR service providers are starting to focus its products and services toward other applications, such as telemetry or remote control.

Churn

Churn is the percentage of customers that discontinue LMR service. Churn is usually expressed as a percentage of existing customers that disconnect over a one-month period. Churn is often the result of natural migration (customers relocating) and switching to other service providers. A majority of the churn for LMR systems is the migration of customers to cellular technology.

Some LMR service providers contribute an activation commission incentive to help reduce the sale price of the radio, and this can be a significant cost if the churn rate is high. Some LMR carriers and their agents have gone to various lengths to reduce churn, including the signing of lengthy service agreements. These service agreements have a penalty fee in the event the customer disconnects service before the end of the period.

Availability of Equipment

The design and production of LMR radio equipment requires significant investment by a manufacturer. LMR radios are more complex, and portable digital wireless radios are ordinarily larger and more expensive. However, digital LMR systems offer standardized products that have relatively large production quantities. This has increased competition and reduced wholesale prices closer to equivalent analog radios (similar size and features).

Distribution and Retail Channels

Products produced by manufacturers are distributed to consumers via several distribution and retail channels. The key types of distribution channels include representatives, wholesalers, specialty retailers, retail stores and direct sales.

Representatives (commonly called "reps" or "agents") are companies that or people who sell LMR or LMR-related products. Representatives may offer products from other companies and they are typically commissioned to sell products in a specific geographic region. Reps are paid a commission based on their sales. In Germany, Indonesia, Malaysia, the Philippines, Poland, Portugal, South Africa, Spain and Turkey, equipment representatives play a major role in recruiting customers [Source: IMTA].

Wholesalers purchase large shipments from manufacturers and normally ship small quantities to retailers. Wholesalers will usually specialize in a particular product group, such as mobile radios and pagers.

Specialty retailers are companies that focus on a particular product category, such as electronics or communications. Specialty retailers know their products well and are able to educate the consumer on services and benefits. These retailers usually get an added premium via a higher sales price for this service.

Some LMR service providers employ a direct sales staff to service large customers. These direct sales experts can offer specialty service pricing programs. The sales staff may be well trained and regularly sell at the customer's location.

Distribution channels are commonly involved in the activation process. The application for LMR service can be a lengthy process, as it may involve the sale and activation of multiple radios and business credit applications.

Consumer advertising programs are not very successful in the business. In fact, most operators report that less than 10% of customers are recruited using this method. This is most likely because LMR service targets the business user and not the consumer.

Market Growth

The first SMR systems became operational in 1974. By 1998, there were over 4.6 million SMR radio customers in the United States [AMTA/Strategis Group 1999 Study]. Of these, approximately 50% are digital customers. Because of the ability of SMR companies to offer Enhanced SMR (ESMR) cellular-like services, this number is expected to increase dramatically.

ESMR services offer some distinct advantages over other wireless services, including limited or non-existent roaming charges and the ability to contact more than one person simultaneously within the network. Enhanced messaging and dispatch services round out the offering.

While not aimed at attracting the attention of low usage consumers, many companies with a mobile workforce are signing up. The Strategis Group estimates a gradual decline in the number of private radio users in the United States as public systems become more competitive. However, the Strategis Group also predicts a gradual increase in the number of LMR radios purchased as companies replace and upgrade older systems. Figure 8.8 shows that the Strategis Group also predicts that by 2004 there will be approximately 15.1 million private radio users in the United States.

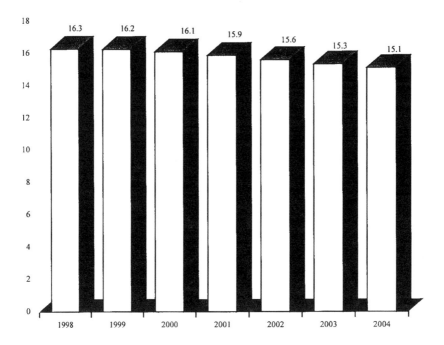

Figure 8.8, SMR and ESMR Market Growth in the United States
Source: Strategis Group, Washington DC

References:
1. Herschel Shosteck Associates, E-STATS. Wheaton, Maryland.
2. Lawrence Harte, Arugs Business, "Techniques," *Cellular Integration Magazine*, January 1996.
3. Interview, Michael Sommer, Information Technologies, October 3rd, 1998.
4. Interview, Bob Glen, Sparton Electronics, Raleigh NC, January 1996.
5. Interview, industry expert, July 15th, 1998.
6. Interview, industry expert, July 15th, 1998.

Chapter 9
Future LMR Technologies

There are several new wireless technologies that will likely influence the LMR industry. These include voice paging, high speed packet data systems, digital channel multicarrier architecture (DC/MA) and spatial division multiple access (SDMA).

Voice Paging

A new development for LMR systems that revises an old application is voice paging. Some of the first paging systems developed in the 1960s used voice paging. These systems were converted to numeric paging due to the inefficiency of sending voice messages.

The difference between the original voice paging systems and advanced voice paging systems is the ability to compress and confirm messages along with the ability to send the messages to the specific radio tower that is serving the mobile radio. The combination of these features allows for the efficient delivery of voice messages. Figure 9.1 shows a typical voice pager.

Traditional voice paging systems transmit a voice message at every radio transmitter site in a geographic service area. Because an average voice message can last 30 to 60 seconds in duration, this is a relatively inefficient method of delivering messages. If the number of transmitters in a geographic area is 100, the sending of each 30 to 60 second message requires 50 to 100 minutes of airtime. This challenge can be overcome by sending the voice page message only to the transmitter where the voice pager is located. Advanced two-way land mobile communication systems have the ability to locate and confirm the delivery of messages to voice pagers.

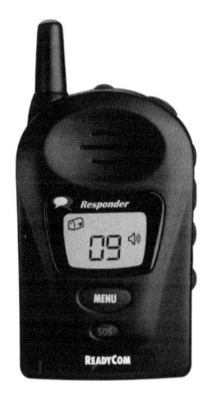

Figure 9.1, Voice Pager
Source: ReadyCom

Figure 9.2 shows the basic operation of a two-way land mobile radio voice messaging system. In the first step, the caller dials a telephone number that is assigned to a voice pager. Once the caller receives a recorded message (step 2), they leave a message that is stored in a voice mailbox. The system then pages the LMR voice pager and attempts to deliver the message (step 3). After answering the page, the voice pager will receive the voice message (step 4) and store it in the voice pagers message memory (step 5). After the message is successfully transferred, the voice pager acknowledges its successful receipt (step 6) and the message is deleted from the voice mail system's memory. The voice pager will then alert the customer that a new message is waiting. The customer can then play the voice message at a convenient time (step 7).

Because land mobile radio systems offer two-way operation, voice paging can provide confirmation of delivery. This allows a land mobile radio system to provide new service, called confirmation paging. Confirmation paging allows the caller to receive a confirmation message after the message has been successfully delivered. The confirmation can be in the form of an automated call back, e-mail or fax message.

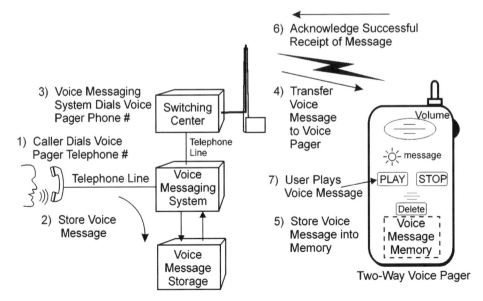

Figure 9.2, Voice Paging System

One of the key advantages to the new voice paging system is the more efficient use of the radio channel through the compression of voice messages. Figure 9.3 shows the basic voice compression process used for voice paging systems. The first step in voice compression is the removal of the pauses between words. Another compression technique is the acceleration of voice playback. Because the message is not sent in real time, it can be sent at a faster transfer rate that allows it to use less time on a radio channel. For example if a voice message that is two minutes long can be accelerated by a factor of two, it will only require one minute to transfer the message. When the message is played back, its speed is slowed back down to the original rate.

In addition to the advantages of compressed voice, non-real time messaging allows for extended battery life. Typical land mobile radios must regularly listen for paging messages. They cannot sleep for more than a few seconds between expected calls (pages) or the caller may hang up. Alternatively, voice pagers can sleep for several minutes and wake up for short periods (tenths of seconds). This can increase battery life by over ten times the battery life of a typical land mobile radio.

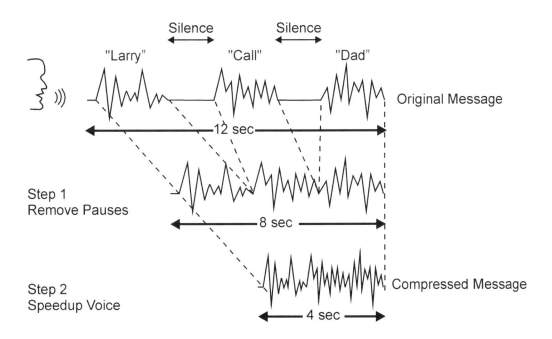

Figure 9.3, Compressing Voice Messages

High Speed Packet Systems

Low speed wireless data systems have been in use since the 1980s. Recently, land mobile radio high-speed packet data services have been demonstrated that offer over 100 kbps data transfer rate. High-speed data services allow land mobile radio systems to support new types of applications that would be unacceptable at low speed data transfer rates.

Figure 9.4 shows an example of a land mobile radio high-speed packet data system. In this diagram, several sub-band data channels are combined to offer high-speed packet data transfer. The high-speed data link (100 kbps) is sent to a demultiplexer. The demultiplexer divides the data channel into four 25 kbps sub-channels. Each sub-channel is supplied to its own sub-channel modulator. The low level modulated signals are then combined, amplified, and transmitted as a 25 kHz radio carrier.

Sub-band data channels are often used in paging systems to divide a single carrier (e.g., 25 kHz) into several sub-channels (e.g., 6.25 kHz each). High-speed packet systems may combine several of these sub-channels to transfer at a much higher data rate.

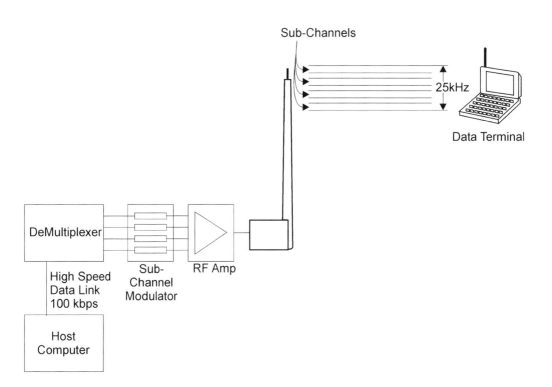

Figure 9.4, High-Speed Packet System

Digital Channel Multicarrier Architecture (DC/MA™)

Digital Channel/Multicarrier Architecture (DC/MA™) allows multiple voice and data carriers to operate simultaneously within the existing regulated bandwidth of a licensed radio channel and is frequency independent. This technology is imbedded in a custom integrated circuit that can be added to a mobile, portable or base station transceiver (or other wireless devices). DC/MA™ is scaleable in 4kHz increments allowing it to be used in different frequency bands and channel bandwidths from 5kHz to 30kHz. Each 4 kHz increment in occupied bandwidth DC/MA™ provides two independent, time multiplexed channels so a 5 kHz (e.g., 220MHz) channel capacity is doubled and a 25kHz (e.g., 800MHz) channel capacity increases eight times.

The DC/MA™ system uses advanced digital signal processing techniques to compress and digitize voice signals into very small frequency slices then time multiplex them to get even more capacity. Additionally the DSP (Digital Signal Processor) must filter both transmit and receive signals to prevent interference to the DC/MA™ transceiver and any adjacent channel systems.

DC/MA™ chips contain a DSP, microcontroller, memory, analog to digital convertors (ADCs), digital to analog convertors (DACs) and the I/O (Input/Output) interfaces needed to control the radio unit, making all models equipped with these devices compatible regardless of manufacturer. This technology provides capacity improvement and a common digital air interface for many types of radio systems. (See Figure 9.5.)

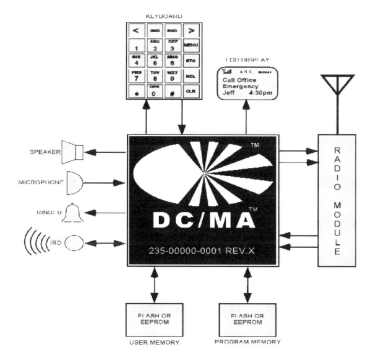

Figure 9.5, DC/MA™ Chip Added To Radio Design
Source: ComSpace

DC/MA™ technology allows land mobile radio operators simple conversion/upgrade options. The conversion involves only changing base station transceivers one channel at a time. All other infrastructure (antennas, combiners, etc.) remains the same. New subscriber sets with the DC/MA™ chip are required for the new base station, but the chip technology is multi-mode, allowing the new units to continue to use and roam to old unconverted base stations until these are ready for upgrading.

DC/MA™ technology can be used as either voice or data channels. Each time divided sub-channel transfers data at 8 kbps and the two sub-channels in each 4kHz frequency segment can be combined for data rates up to 16 kbps. Using a feature called dynamic channel allocation, multiple frequency segments can be combined for even higher data rates.

The DC/MA™ system can be used for simplex or duplex operation (with improved capacity). Also since DC/MA™ is digital, it can use Time Division Duplexing (TDD), allowing full duplex operation (even closely spaced frequencies, i.e., 220MHz) without the need of expensive and bulky duplex filters. It accepts most trunking protocols including LTR®, LTR Netä, ESASä, Passportä, MPT 1327, etc. A feature called Service Option Negotiation ("SON") translates each protocol into a high-speed digital message that can be transparently added to each sub-channel.

Spatial Division Multiple Access (SDMA)

Spatial Division Multiple Access (SDMA) uses narrow focused radio coverage areas to increase the capacity and quality of a system. Using advanced antenna systems and digital signal processing, it is possible for a base station to focus its transmission to a specific mobile radio. This increases the effective power that is transmitted to the mobile radio and helps to reject interference from nearby transmitters that are not directed toward the focused radio coverage area. The benefits of SDMA technology include larger radio coverage areas (less radio sites in a system), greater ability to receive signals from low power portable mobile radios and the ability to have more channels in a single radio site (greater network capacity). SDMA technology can be applied to any wireless access systems (FDMA, TDMA, CDMA, etc.).

Figure 9.6 shows a typical 120 degree sectored radio site (an antenna which covers 1/3 of the radio site area) as compared to a radio site that uses SDMA technology. In example (A), a mobile radio communicates with the radio site anywhere in the coverage area on frequency 1. The base station is transmitting 10 Watts which is distributed over the entire third of the radio site. Only a small amount of this RF energy actually is received by the mobile radio. In example (B), a larger geographic area can be covered using the same amount of RF energy and multiple channels on the same frequency can co-exist. By focusing 10 watts of energy to a specific region, a majority of RF energy reaches the land mobile radio. This allows the mobile radio to operate at a further distance from the radio site. Additionally, multiple focused radio beams can be created to allow a single radio site to simultaneously service more customers.

An SDMA equipped radio site will continually track the position and movement of each mobile radio in its coverage area. As the mobile radio moves, the radio site will change the direction of the radio coverage beam to provide coverage to the specific mobile radio. If a mobile radio is operating on the same frequency as another mobile radio being serviced by the radio site and moves too close to a third mobile radio, the system will automatically change the frequency so they do not interfere with each other.

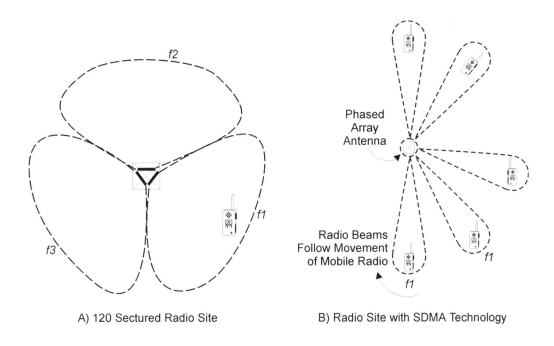

A) 120 Sectured Radio Site   B) Radio Site with SDMA Technology

Figure 9.6, Spatial Division Multiple Access (SDMA) Technology

Dual Technology Mobile Radios

Some land mobile radio systems have been designed to allow dual mode operation (use of multiple types of channels). If a mobile radio can receive only on a single analog or digital technology, they can only operate in that specific type of digital service. Some users that only require service in a small geographic area (such as taxi cabs or local delivery services) may find this acceptable. However, users that require radio service in a wide geographic area where a mix of old and new technologies exist can benefit significantly from the availability of dual mode radios. Dual mode operation involves the use of a single control channel to assign access to the old and new types of radio channels.

It is possible to combine other wireless technologies with land mobile radios for dual mode service. This includes dual mode cellular service with LMR radio technology (e.g., GSM and MPT1327) or satellite service with LMR radios (e.g., Iridium and iDEN®). The benefits of dual technology mobile radios include advanced services and international roaming.

Chapter 10

Advanced LMR Services

Almost all LMR systems are capable of providing advanced features. Advanced features may be provided by the LMR network (e.g., voice mail), by the mobile radio (e.g., speed dialing) or a combination of the mobile radio and the system (e.g., short messaging).

The new digital LMR systems are capable of providing many types of voice, messaging, and data services. Most of these services are compatible with services offered by other types of networks (typically the public telephone network). To ensure these services can correctly operate with each other, LMR specifications detail many of the precise operational parts of these services and how they are expected to interact with other services.

Short Messaging

Short messaging service provides text information to one or more mobile radios. Some short message services may be paid for by customers, while other services may be paid for by merchants. Short message services can be delivered by data messages that appear on the display or by audio messages. Examples of some short message services include weather and traffic reports, imaging transfer and direction routing/maps.

Separate messaging services can be designated for direct and indirect human applications to demonstrate the existing and the potential market for wireless information services. Direct human message services include paging, news and traffic reporting, direction routing, etc. Indirect human uses include monitoring and remote control of machines, such as electric machines and water meters, gas valves and thousands of other types of devices. Indirect human applications of wireless messaging appear to have a higher potential number of uses than human applications.

Short messages usually contain about one page of text, or approximately two thousand bytes of information or less. Some systems limit the short message size to approximately 200 alphanumeric characters. Short messages may be received while the mobile radio is in standby (idle) or in use (conversation). While the mobile radio is communicating both voice and message information, short message transfer takes slightly longer while the mobile radio is in standby.

Short message service (SMS) can be divided into three general categories: Point-to-point, point-to-multi-point and broadcast. Point-to-point SMS sends a message to a single receiver. Point-to-multi-point SMS sends a message to several receivers. Broadcast SMS sends the same message to all receivers in a given area. Broadcast SMS differs from point-to-multi-point because it places a unique "address" with the message to be received. Only mobile radios capable of decoding that address receive the message.

Point-to-Point Messaging

Much like voice mail or beeper type systems, a point-to-point message system sends a short message from one source to one receiver. Figure 10.1 illustrates point-to-point SMS transfer. Initially, the message is received by a dispatcher and is routed to the land mobile network (step 1) where it is first routed and stored in a message service center (step 2.) The LMR network then searches for the mobile radio (step 3) and alerts the mobile radio that a message is coming. The mobile radio tunes to the radio channel where the message will be sent. The system then attempts to send the message (step 4). As the message is being sent, the system waits for acknowledgment messages (step 5) to confirm accurate delivery of each part of the message. If the transmission is successful, the message may be removed from message center memory. If unsuccessful, the system attempts delivery again.

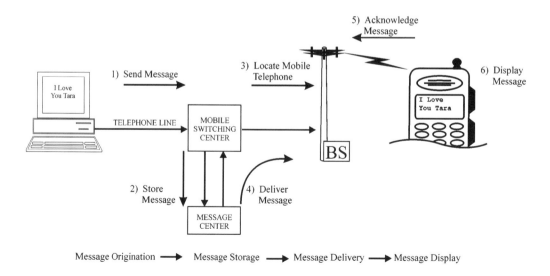

Figure 10.1, Point-to-Point Messaging

· Messages can be sent to mobile radios (called mobile terminated messages) or messages can be sent from mobile radios (called mobile originated messages). Because mobile radios usually have a very limited number of keys (usually used to dial calls) it is difficult to key in (originate) messages. The result of this is that a majority of messages are sent from other sources (e.g., computer console) to mobile radios. However, some mobile radios have the option to connect to external keyboards or telemetry monitoring that allows the user to easily enter text messages. Another common alternative used in LMR systems is the use of predefined messages that can be selected from a list by the customer. This allows easy sending of short messages from mobile radios with a limited number of keys.

Point-to-Multi-Point Messaging

Point-to-multi-point messaging is a process of sending the same message to a predefined group of mobile radios. Figure 10.2 illustrates how a message is transferred to a group of users through the use of point-to-multi-point messaging. Similar to point-to-point message service, a message is first routed to a message center (step 1). When the message is received by the message center, it is determined that the message is designated for multiple mobile radios. The designated list of message recipients may be pre-arranged (such as a sales staff) or it may be included with the originating message. In either case, the message center stores the message and recipient list (step 2).

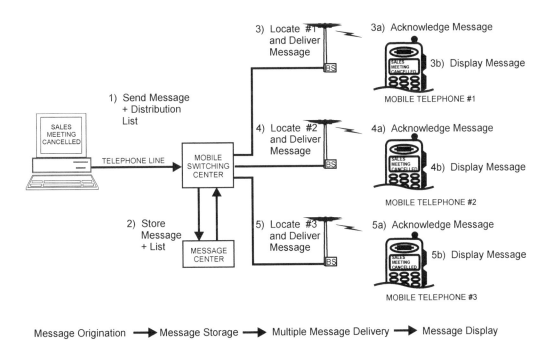

Figure 10.2, Point-to-Multi-Point Messaging

To complete message delivery, the system then searches for each mobile radio in the list. The LMR system then individually delivers the message to each mobile radio (steps 3-5). To deliver the message, the LMR system alerts (pages) the mobile radio. The alert message contains an indication that the alert message is for the delivery of a short message. The mobile radios then tune to the designated radio channel (voice or control), alerts the system that it is ready to receive a short message and the system sends the message. If the transmission is successful, the mobile radio ordinarily sends an acknowledgment, and the message may be removed from short message center memory. If the message transmission to one or more of the receivers was not successful, the LMR system may attempt delivery again later. An example of point-to-multipoint messaging might be a message to a corporate sales team indicating "Sales meeting on Tuesday is canceled."

Broadcast Messaging

Broadcast messaging involves the sending of a message to all mobile radios that have been pre-set to monitor a specific broadcast message channel. Every base station transmits each broadcast message along with enough information to accurately decode the message. Broadcast messaging sends a single message to a large geographic area, and all the mobile radios equipped to receive the message simultaneously decode the message.

Figure 10.3 illustrates how broadcast messages are delivered to users. Like the point-to-point message service, the message first goes to the message center (step 1) where the message center determines that the message is designated for all mobile radios with a unique broadcast code. The delivery code may be pre-arranged (such as an advertising code) or included with the

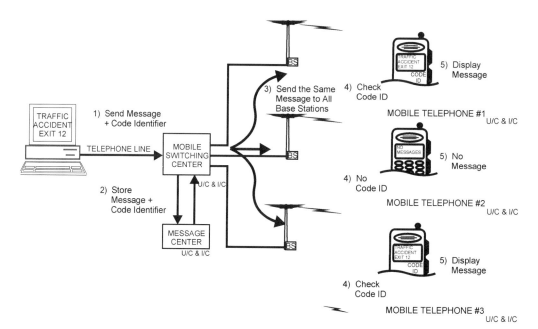

Figure 10.3, Broadcast Messaging

originating message. The message is stored with the code identifier (step 2). The message then broadcasts on a designated message channel (step 3), which may be part of a control channel in every cell site where the broadcast message is to be transmitted.

Unlike point-to-point and point-to-multi-point messages, mobile radios do not acknowledge receipt of broadcast messages. If a mobile radio is off, or is not tuned to the message channel, it misses the broadcast message. To address this limitation, messages may be broadcast several times, and mobile radios that have already received the message may ignore repeats (steps 4-5).

An example of a service that would use broadcast messaging is advertising or weather reporting. LMR service providers could send advertising or projected weather conditions to individuals or groups of customers.

Executable Messages

Executable messages are programs that are sent via short message and are received by a mobile radio and that instruct the mobile radio to perform processing instructions. Examples of executable messages include telephone numbers and feature programming.

Figure 10.4 shows how an executable message is sent to an LMR mobile phone. In this example, the LMR system operator (dispatcher) desires to send a new feature to a customer (such as a speed dial telephone number for customer service). In this example, a dispatch center computer sends the message through the LMR system where it is received by the phone and stored in the mobile radio. The microprocessor in the mobile radio card is used to process the message.

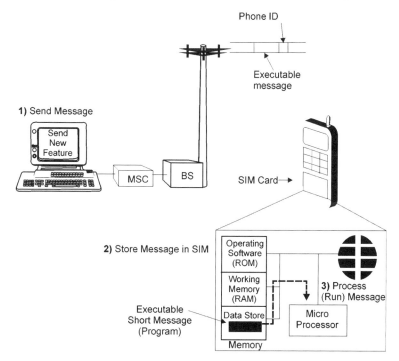

Figure 10.4, Executable Short Messaging

Data Communication

Wireless data capability is becoming an important part of the LMR service providers' services. Data communications service is often the key that keeps a customer from converting from LMR to other wireless services, such as cellular or paging.

Data communications services include continuous data, packet data, fax delivery, remote control, and telemetry/monitoring. Circuit switched data provides for continuous data signals, packet switched data allows for very short data messages and fax service allows the verification that complete fax messages are received without errors from radio transmission. A common industry term for monitoring and remote control through LMR systems is supervisory control and data acquisition (SCADA).

Circuit Switched Data

Circuit switched data is a temporary dedicated circuit between two points within a network. The circuit is temporary since it is only used for the duration of a call and it is unshared by other users. Once the data transmission session is complete, the circuit is disconnected and the radio or voice channel becomes available for other users.

Circuit switched connections are efficient for data transmission that requires the continuous sending of data. This includes fax transmissions or file transfers where data is being exchanged for a high proportion of the connection time.

Circuit switched data can be sent on analog or digital radio channels. When sending data on an analog channel, a data modem is required. A modem converts digital information to audio tones that can be sent through the telephone network. When transmitting data on a digital channel, a data adapter is required. The data adapter temporarily stores digital information and adapts the data transfer rate of the radio channel. After digital information is received by a LMR system, it is usually converted by a modem so the digital signal can be transferred through the public telephone network.

Analog radio channels normally transport data from 2400 to 19200 bits per second, and digital radio channels usually transfer at slightly higher data rates. The data transfer rate is greatly dependent on the quality of the radio signal.

Figure 10.5 shows how an LMR system sends circuit switched data over an analog radio channel. In this example, a laptop computer is sending a file to a computer in an office. The laptop computer is connected to modem, and the modem is connected to an LMR data capable mobile radio.

To initiate data transmission through the system, the laptop computer sends a command to initiate a call via the modem to the LMR. This command includes the dialed digits of the destination phone number of the office computer. The modem then initiates the mobile radio to send a service request to the LMR system along with the dialed digits. The LMR switching center then dials the telephone number of the destination computer and the modems begin to communicate with each other. After the modems have completed their initialization (called training) and adapted their data transfer rates, the data connection between the laptop computer and office computer is complete and the file transfer begins.

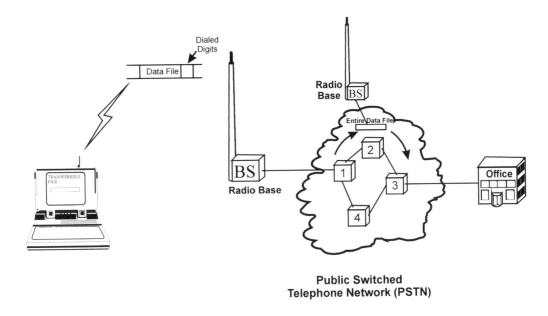

Figure 10.5, Circuit Switched Data Services on an Analog Radio Channel

To enhance the data transfer rate on analog systems, several special modem protocols have been developed. These protocols differ from standard modem protocols, as they can adapt to the distortions that occur on the radio channels. These protocols include Microcom Networking Protocol Class 10 (MNP10) and Enhanced Throughput Cellular (ETC). The use of modem protocols, specifically designed for radio transmission, increase the reliability of the data connection by being able to adapt to the varying radio channel quality.

Radio channels on digital LMR systems were designed to efficiently transfer data. However, audio signals for digital LMR systems are compressed by a speech coder that limits the ability to transfer data. Speech coders do not effectively compress standard modem signals. As a result, it is not usually possible to send data or fax information directly through the audio section of a digital LMR radio.

In order to send data information through a digital LMR, the data signal must bypass the speech coding process. The data connection between the computing equipment and the mobile phone does not require a modem. Instead, the connection only requires a data adapter. The data adapter primarily buffers (isolates) and temporarily stores information that is to be transferred.

After the data from the computer is sent through the data adapter, the digital mobile radio adds error protection bits to the digital signal that is to be transmitted. The amount of error protection bits added depends on the type of device that is connected and the desires of the user. For example, for a file transfer, the amount of error protection may be very high. For a slow scan video signal, the error protection may be very low as a slightly distorted digital picture is acceptable.

Figure 10.6 shows how a LMR system sends circuit switched data over a digital radio channel. In this example, a laptop computer is sending a file to an office computer. The laptop computer is connected to a data adapter and the data adapter is connected to a digital LMR that has data transfer capability.

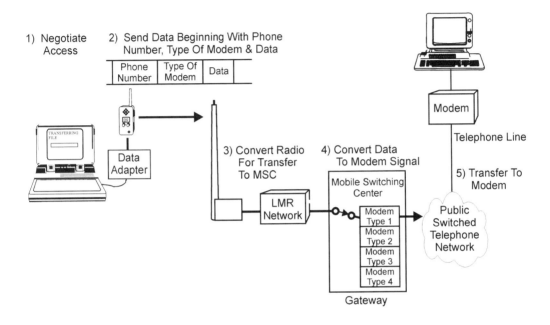

Figure 10.6, Circuit Switched Data Services on a Digital Radio Channel

To initiate data transmission through the system, the laptop computer sends a request to send data along with the desired destination telephone number. This request is transferred through the data adapter to the digital LMR. The digital LMR then sends a service request indicating the destination phone number along with a request that the call is a data call. When the LMR system receives the request, it connects the call to one of several modems in the mobile switching center. These modems are the inter-working function (IWF) gateway that converts the radio data into a modem signal that the landline modem can understand. The LMR switching center then dials the telephone number of the destination computer and the modems begin to communicate with each other. After the modems have completed training (synchronizing their signals), the data connection between the data transfer, the laptop computer and office computer may commence.

Packet Switched Data

Packet Radio Service provides transfer of small packets of data between mobile radios and data networks (such as the Internet). Most digital LMR networks were designed for continuous data transfer for voice communication or continuous modem data transfer. The design requirements for voice transmission, particularly the long call setup requirements, limit the ability of digital LMR networks to effectively offer packet data services. However, some digital LMR systems have been designed to offer both voice and packet data services.

Circuit switched and packet switched data services are very different. Circuit switched service is typically used for a single lengthy transfer of data while packet service is used for rapid transfer of small packets of information. Circuit switched data can accept a lengthy call setup time of a few seconds, while packet data transmission requires rapid channel assignment and tear down. Because circuit switched data transfer has a relatively large amount of call setup and tear down, it is ordinarily inefficient for serving sensing control and applications that require small amounts of information that must be sent very quickly. Packet radio service is best suited for many of these applications that require small amounts of data when communicating.

Some LMR systems already offer packet radio service. Typical applications for packet radio service include telemetry (e.g., meter monitoring), wireless e-mail, train control systems and credit card processing. Typical packet sizes range from 100 to 1500 bytes of data.

Rate charging for packet transaction is usually based on the volume of data the user expects to transfer. This means that a device could be connected for an extended period of time and the user is only charged for the actual data transmitted. Typical rate plans are flat rate with a user entitled to a certain number of kilobytes of data. When the user exceeds the entitlement, additional data is charged at a fixed rate per kilobyte.

Figure 10.7 shows a packet switched data service. In this diagram, a wireless credit card machine is connected to a data adapter and a digital LMR. After a credit card is swiped, the machine sends approximately 100 bytes of information to the data adapter. The data adapter divides the information into two parts (packets). Each packet is given its destination address and

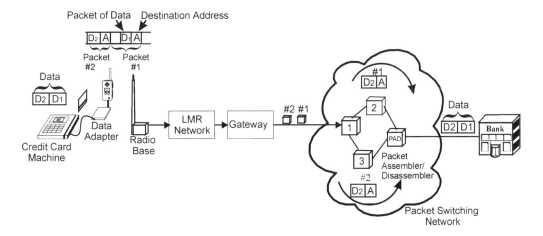

Figure 10.7, Packet Data Service

a sequence number (#1 and #2). The LMR sends each packet of data to the LMR system. When the LMR system receives the packets, it passes them onto a gateway. The gateway adapts the packets (adjusts the timing and levels) for transmission through a packet switching network (such as the Internet). Each packet travels on the best possible route to its destination. This diagram shows that packet #1 is successfully sent through switch number 2 to its destination while packet #2 is sent through switch number 3. This occurs because switch number 2 was busy at the time of transmission request. The destination is a packet assembler/disassembler (PAD). The PAD places the received packets in correct order and removes the address information so the original data message can be recreated. After the message is received and decoded by the bank's computer, the credit card processing continues.

Fax Delivery

Facsimile (fax) delivery is the transfer of scanned or pre-stored information from one location to another. This information includes purchase orders, invoices, brochures, or any other supporting documentation. Remote fax machines can be connected to regular landline phones, mobile radios and laptop computers. Fax machines are now beginning to appear in airports, hotels, conference centers and business centers making remote fax delivery more convenient.

Fax transmissions are similar to those of standard data modem transmissions. However, while modem data transfer algorithms have been adapted to work well with landline telephone systems, poor radio connections limit the use in LMR systems. Fax transmission schemes do not recover or correct transmission errors well. Depending on the error type and location, some or all of a wireless fax page can be lost or the fax page may distorted with black or white streaks as a result of data transmission errors. With newer digital data transmission protocols, this need not be the case. Digital data transmission protocols correct errors through an encoding scheme and through re-transmission of missing or incorrect data. Assuming there are no dropped calls, wireless fax transmissions could be as reliable as a wire line.

Figure 10.8 shows a portable fax display device that can be connected to a mobile radio. When the faxed information is received, the mobile radio routes the audio signal to the fax receiver in the display device. The information is then stored in the memory of the fax display. To view the fax, the customer looks into the display unit and the information appears. Simple menu control and zoom features are provided to allow the customer to navigate through multiple pages. The menu can be used to forward a fax or send a new one with a text message.

Figure 10.8, Portable Fax Display Device

In cases where fax delivery is attempted and the mobile radio is either not available or appears in a poor radio coverage area, some wireless systems can store faxed messages and forward them when delivery becomes possible.

Telemetry/Monitoring

Wireless service provides a way to transfer telemetry and monitoring information without wire line connections. Telemetry applications include monitoring utility meters, gas lines, vending machines, critical equipment, environmental sensors (water level, earthquake or fire) and many others. For applications that transmit only small amounts of information, a packet transmission system may be cost-effective. When monitoring devices require the transfer of large amounts of data (such as a security camera), circuit switched data transfer is better suited. Figure 10.9 shows an electric meter that can be monitored remotely via wireless service.

Figure 10.9, Electric Meter with Wireless Telemetry Capability
Source: ABB

Remote Control

In distant and rural areas, it is often impractical to install telephone or remote control lines. Radio receiver modules can be installed to control devices such as gates, vents, water valves, gas valves, audio alert devices (e.g., sirens), and visual displays (e.g., road signs). If power is not available to control these devices, power could be obtained from batteries, solar panels or even from a generator that runs from the flow of gas or water. Remote control service has been used by LMR systems for over two decades. Figure 10.10 shows a remote control device that has the ability to disconnect an electric appliance (commonly an electric water heater) via radio remote control.

Image Services

Mobile radios can reliably send and receive pictures from digital cameras or scanners. These images can be sent to any place in the world through standard telephone lines or other networks such as the Internet. Image services use digital data transfer to physically transport information between devices and applications (such as video data compression). Image services include photo transfer and video transfer.

Figure 10.10, Wireless Remote
Control Device
Source: EchoPort, Inc

Photo Transfer

Digital cameras allow visual information to be captured, transferred, and stored without film almost instantly and efficiently. Insurance agents, security services, reporters and police, among others, use still image capture devices. The benefits for a wireless digital image transfer system include rapid information capture (e.g., on-site insurance adjustment), instant distribution, and the lack of a film, film developing, postage or physical picture storage requirement.

Radio channels occasionally distort signals and cause file transfer errors that could distort the images. Image data files that must be transferred will benefit from a reliable direct digital channel. Unfortunately, direct digital channels connect only the mobile radio and the wireless system. For data to reach a remote computer, it must be converted between the wireless system and the PSTN.

Figure 10.11 shows a wireless image capture system. After the user selects the appropriate viewpoint, the user presses the transmit button. The digital camera captures a digital image and the image is digitally compressed and then supplied to the mobile radio for digital transmission. The mobile radio sets up a call to a remote computer where it receives the picture file that can be printed, viewed on a display or stored for subsequent retrieval.

Video Transfer

Although some LMR systems transfer real time video images, the limited amount of data transfer typically restricts the sending of video to low resolution slow scan video. Newscasters sometimes use non-real time video transfer to send high-resolution video over a radio channel.

Because high-resolution video clips are comprised of large amounts of information, they ordinarily cannot be sent over a single radio channel at the same time they are recorded. When sending a high-resolution video clip via a single wireless channel, it is

Figure 10.11, RADCAM Image Capture System
Source: Kenwood Communications

sent in non-real time. The video signal is first digitized, then compressed and sent as a large data file. Even with a significant amount of data compression, the data file will take longer to send than the recording time. When the data file is received, it is uncompressed and can be played back at the same rate that it was recorded.

Video transfer systems can be used with an LMR system. To accomplish this, a digital video camera records a news event. This digital video is stored in a file and compressed over 100:1. The compressed video data file is then sent to a data modem on its way to an LMR mobile unit. The LMR mobile transfers the file to the LMR network that routes the file to a computer in the news office. This computer decompresses the file so it can be edited for television.

Location Monitoring

Location monitoring is the process of tracking the position of mobile radios. Location monitoring is used for vehicle tracking (such as trucks), survey site locations, security (i.e., personal location), inventory and asset (e.g., equipment) management and a variety of other functions.

There are two methods (or a combination of the two) used for location monitoring. The first method uses the system to determine the position of the mobile radio. The second method uses external position location devices to track the location of the mobile radio.

System Position Location

Most new digital wireless systems have built-in potential for basic location monitoring. Mobile radios can be quickly located to the radio site in which they are operating. This normally can locate the mobile radio within an approximate 15 km (12 mile) radius. While this may be acceptable for long-haul trucking applications, it is often necessary to locate the mobile radio within a hundred meters. In 1996, the FCC designated that commercial mobile radios must be capable of being located within 125 meters by the year 2002.

Almost all digital systems have some ability to provide more precise location information on mobile radios operating in the system. These systems must usually compensate for the mobile radios distance from the radio site to adjust for transmission time delays. This built-in distance monitoring makes it likely that future digital systems will offer some mobile radio location services.

Figure 10.12 shows a sample position location system provided by the network. The very first step in the position location system is to determine in what radio site area the mobile radio is operating. The radio site identification can locate a mobile radio within a few miles. Most LMR systems already have this capability. By using a directional antenna system or by monitoring the same signal on nearby antennas (called triangulation), the relative angle (and possibly position) from the radio site can be determined. If the LMR system has the capability to track the transmission delay, an approximate distance from the radio site antenna can also be used to help pinpoint the location of the mobile radio.

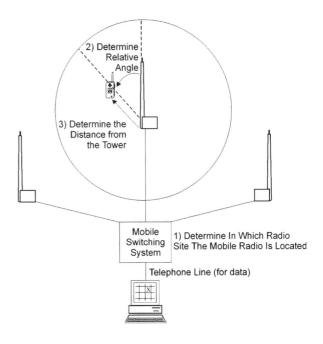

Figure 10.12, Network Position Location System

External Position Location

Another way to monitor a location is to attach a Global Positioning Satellite (GPS) receiver to a mobile radio. The GPS device determines its position from satellites orbiting the earth. After the GPS unit has determined its location, the mobile radio can transfer the position data to a computer in the wireless system via the radio channel.

Figure 10.13 shows a position location system that uses an external position location GPS receiver that attaches to a land mobile radio. As the mobile radio moves throughout the system, the GPS receiver continuously calculates the position of the mobile radio. The location information can be continually sent through the system or it can be requested on command by the dispatch or switching center.

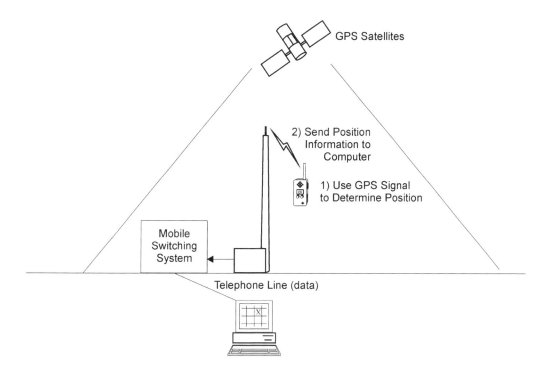

Figure 10.13, External Device Position Location System

Dispatch

Dispatch systems allow the coordination of groups of users through the use of simultaneous broadcasting of messages to a group of users. An LMR system has the ability to provide for voice and computer-aided dispatch service.

Voice Dispatch

Dispatch voice service allows for the simultaneous reception of speech conversation of a pre-defined group of mobile radios. Each mobile radio that has group-call capability is called a group call member. A dispatch console can be connected to an LMR network via wires or by radio channel coordination.

Dispatch service usually operates in half-duplex mode, which has the ability to transfer voice or data information in either direction between communications devices but not at the same time. This means that two-way conversation is not in progress during a voice group-call. Only the originating group caller is transmitting while all the other mobile radios in the group are listening to the message that follows the group ID code. A dispatcher is a coordinator for group ID. This means a dispatcher can typically hear and broadcast to all group ID members.

Each user in a group is assigned a group ID code. A single mobile radio or dispatcher can have more than one group code. Group calls can be initiated by a customer (with group call capability identification) or by a dispatcher who can listen to all and transmit to all mobile radios within the radio coverage area.

Dispatch calls can be limited to a specific geographic area or to the entire coverage area of the LMR network operator. In some cases, dispatch calls may be patched to other networks. Applications for group call services commonly involve multiple group members in a small service area. The calling mobile radios, as well as the destination mobile radios, may be any compatible mobile radio that has subscribed to the related group ID. Group call messages are sent to geographic areas, where mobile radios receive the message by using the group ID code. Destination mobile radios shall be notified with the group ID, not by paging the mobile radio individually. Mobile radios individually respond to dispatchers by their unique identification code. The group-call feature may cover a service area that is comprised of one or many radio sites. Service areas are predefined in the network by the LMR service provider and coordinated by the network operator.

Dispatchers usually have priority to talk at any moment without a need to signal the desire to talk. Mobile radios usually have to indicate when they want to talk. They are only able to become talking users if there is no other mobile radio user talking. The right to be a talking mobile radio user is often allocated based on the first service request received without queuing. The service customer wishing to talk will usually be given an audible indication that speech can begin (a short beep). After a mobile radio user finishes talking (typically by releasing the push-to-talk button), the mobile radio automatically switches to the listening mode again. Authentication and confidentiality for mobile radio dispatch services are usually optional.

In some cases, several dispatchers may be talking simultaneously. Dispatchers should hear all combinations of voices other than their own. In these cases, group users who are listening hear the combination of all voices.

Different levels of priority for dispatch service may be applied. On the digital LMR systems, a number of group calls may exist simultaneously intended for different groups of destination users in the same service area. Parallel group calls are possible to the same group of destination customers in different (possibly overlapping) service areas. A group call shall be released on demand by the calling mobile radio, by a dispatcher or by the network.

For certain levels of priorities, an acknowledgment of receipt of a group call can be required as an application option (e.g., for railway emergency calls) from all or from designated mobile radios. The acknowledgment itself shall be performed at the end of the dispatch call. The acknowledgment indicates the time the reception started and the time the reception terminated. The acknowledgment has to be given to a pre-defined recipient. It is possible for a dispatch group customer to activate or deactivate the group call reception for different group IDs. The selection list is stored in the mobile radio that corresponds with the subscribed group IDs. It is possible to prohibit the deactivation of group IDs used for high priority calls.

Voice dispatch systems allow a single user to communicate with a group of mobile radios. The group originator waits until a group channel is clear (other group users are not talking on the group channel). When the user presses the talk button (group call), this sends a request message to the system to initiate a group call along with the identification of the phone. This is sent to the group call register where the call will be connected to base stations and/or dispatchers that belong to the group. The group call is then patched or dispatched to all the active users in the group.

Computer Aided Dispatch

Computer aided dispatch systems are often used for the automatic delivery of messages or instructions. A computer aided dispatch system may be capable of monitoring the position of mobile radios and sending messages to units located within a specific area. For example, a CAD system used by the police may send trouble messages to police cars that are in an area near the incident.

Appendix I - Acronyms

A/D - Analog to Digital
AC - Authentication Center (also AuC) or Administration Center
ACCH - Associated Control Channel
ACCOLC - Access Overload Class
ACIPR - Adjacent Carrier Interference Protection Ratio, i.e., the ratio of the strongest adjacent carrier power (compared to the desired carrier power) that can be present at the receiving antenna without degrading the desired carrier signal
ACK - Acknowledge
ADPCM - Adaptive Differential Pulse Code Modulation
AGC - Automatic Gain Control
AGCH - Access Grant CHannel
AGRAS - Air- Ground Radiotelephone Automated Service
AGRASCCN - Air- Ground Radiotelephone Automated Service Credit Card Number
Akey - Authentication key
ALEC - Alternative Local Exchange Carrier
AM - (1) Amplitude Modulation; (2) Ante Meridian, i.e., before noon
AMA - Automatic Message Accounting
AMPS - Advanced Mobile Phone System
ANI - Automatic Number Identifier
APC - Adaptive Predictive Correction speech codec
APCO - Associated Public safety Communications Officers
ARFCN - Absolute RF Channel Number
ARQ - Automatic ReQuest to retransmit, a protocol in which a packet or block of information will automatically be retransmitted within a pre-specified time interval if not acknowledged
ARTS - Advanced Radio Technology Subcommittee
ASCII - American Standard Code for Information Interchange
ASIC - Application Specific Integrated Circuit
ASK - Amplitude Shift Keying
ATIS - Alliance for Telecommunications Industry Solutions
AuC - Authentication Center (database), associated with HLR
AVL - Automatic Vehicle Location
BAIC - Barring of All Incoming Calls
BAOC - Barring of All Outgoing Calls
BCCH - Broadcast Control CHannel
B-CDMA - Broadband CDMA
BCH - Broadcast CHannel

BER - Bit Error Rate, the ratio of erroneously received bits to total received bits

BETRS - Basic Exchange Telecommunications (via) Radio Service; rural radio telephone service

BOC - Bell Operating Company

BTA - Basic (rural, suburban) Trading Area

BTN - Billing Telephone Number

BVA - Billing Validation Application

C/N - Carrier to Noise ratio

C7 - Common channel signaling system no. 7, alternate abbreviation for CCS7

CAD - Computer Aided Dispatch

CAF - Cellular Anti-Fraud

CAP - Competitive Access Provider

CATV - CAble TeleVision

CBCH - Cell Broadcast CHannel (GSM and related systems)

CC - Call Control

CCH - Common Control Channel

CCITT - International Telegraph and Telephone Consultative Committee (succeeded by ITU-T)

CCS7 - Common Channel Signaling system number 7

CDG - CDMA Development Group

CDMA - Code Division Multiple Access

CDPD - Cellular Digital Packet Data

CEC - Commission of the European Communities

CEPT - Conférence Européenne (des Administrations) des Postes et des Télécommunications, i.e., European Conference of Posts and Telecommunications (standards activities succeeded by ETSI)

CFR - Code of Federal Regulations

CFU - Call Forwarding Unconditional

CGSA - Cellular Geographic Service Area

CLID - Calling Line IDentification; caller ID.

CLLI - Common Language Location Identifier

CMA - Communications Managers Association

CMRS - Commercial Mobile Radio Service

CP - Cellular Provider

CPA - Combined Paging and Access

CPC - Certificate of Public Convenience

CPE - Cellular Provider Equipment

CPP - Calling Party Pays

CQM - Channel Quality Measurement

CRC - Cyclic Redundancy Code

CSMA - Carrier Sense Multiple Access

CSS - Cellular Subscriber Station

CT(n) - Cordless Telephony (n^{th} generation)

CT-2 - Cordless Telephone 2nd generation

CT-2+ - Cordless telephone product trade name used by Nortel

CT-3 - Cordless telephone product trade name used by Ericsson Radio Systems

CT1 - Cordless Telephone - 1st Generation

CT2 - Cellular Telephone - 2nd Generation
CTCSS - Continuous Tone Coded Squelch System
CTI - Computer Telephony Integration
CTIA - Cellular Telecommunications Industry Association
CTRC - Canadian Television and Radio Commission (successor to DOC)
CWTA - Canadian Wireless Telecommunications Association
D/R - Distance to cell Radius ratio
D-AMPS - Dual-mode AMPS
DAC - Digital to Analog Converter
DACC - Directory Assistance Call Completion
DAMPS - see D-AMPS
DAP - Dispatch Applications Processor
dB - decibel, a logarithmic representation of the ratio of two power values (named for Alexander Graham Bell)
DCCH - Dedicated Control CHannel
DCS - Digital Cellular System
DCS-1800 - Digital Communications System on 1800 MHz band, UK up-banded and lower power version of GSM
DCSS - Digital Coded Squelch System
DCT - Digital Cordless Telephone, North American version of DECT
DCT-1800 - Digital Cellular Communications-1800
DID - Direct Inward Dialing
Dm - Data mobile channel
DMH - Data Message Handler
DRx - Discontinuous Receive
DSI - Digital Speech Interpolation
DSP - Digital Signal Processing
DSRS - Digital Status Reporting System
DTMF - Dual-Tone Multi-frequency
DTx - Discontinuous Transmission
E2PROM (also EEPROM) - Electrically Eraseable Programmable Read Only Memory
ECB - Enhanced Cordless Base (station)
EDACS - Enhanced Digital Access Communications System
E-GSM - Extended GSM (spectrum)
EIA - Electronics Industries Association
EIA-553 - industry standard for the AMPS cellular system
EMI - ElectroMagnetic Interference
ERP - Effective Radiated Power
ESMR - Enhanced Specialized Mobile Radio
ESN - Electronic Serial Number
E-TACS - European (British, UK) TACS system, see TACS
E-TDMA - Enhanced TDMA, utilizing DSI (Hughes' Network Systems)
ETSI - European Telecommunications Standards Institute (successor to standards activities of CEPT)
FCC - Federal Communications Commission
FCCH - Frequency Correction CHannel (GSM and related systems)

FDD - Frequency Division Duplex
FDM - Frequency Division Multiplexing
FDMA - Frequency Division Multiple Access
FH - Frequency Hopping
FHMA - Frequency Hopping Multiple Access
FIRSTCHA - First access CHAnnel
FIRSTCHP - First paging CHAnnel
FM - Frequency Modulation
FOCC - FOrward analog Control Channel
FPLMTS - Future Public Land Mobile Telephone System
FSK - Frequency Shift Keying
FVC - Forward analog Voice Channel
GDN - Group Dispatch Number
GEO - Geostationary Earth Orbit
GHz - GigaHertz, a thousand million cycles per second
GOS, GoS - Grade of Service
GPS - Global Positioning System
GSA - Geographical Service Area
GSC - Golay Sequential Coding
GSM - Global System for Mobile communication (formerly Groupe Spècial Mobile)
GSM MS - GSM Mobile Station
GSM PLMN - GSM Public Land Mobile Network
HDSL - High Density Subscriber Loop
HF - High Frequency
HLR - Home Location Register
HPLMN - Home Public Lands Mobile Network
HPU - Hand Portable Unit
HT - Hilly Terrain
Hz - Hertz, one frequency unit equal to one cycle per second
IBCN - Integrated Broadband Communications Network
iDEN - integrated Dispatch Enhanced Network
I-ETS - Interim European Telecommunications Standards
IC - Interexchange Carrier
ICA - International Communications Association
IDLC - Integrated Loop Digital Carrier
IDN - Integrated Digital Network
IEC - Inter Exchange Carrier
IMTA - International Mobile Telecommunications Association
IMTS - Improved Mobile Telephone Service
IPR - Intellectual Property Rights
IS - Interim Standard
IS-136 - Interim Standard 136 for North American TDMA cellular system
IS-41 - Interim Standard 41 for North American inter-switch signaling
IS-54 - Interim Standard 54 for the first North American dual-mode TDMA cellular system
IS-88 - Interim Standard 88 for the narrowband AMPS cellular system
IS-95 - Interim Standard 95 for CDMA cellular service.

ISDN - Integrated Services Digital Network
ISDN-BRI - Integrated Services Digital Network - Basic Rate Interface
ISDN-PRI - Integrated Services Digital Network - Primary Rate Interface
ISDN-UP - Integrated Services Digital Network - User Part
ISO - International Standards Organization
ITA - Indiana Telecommunications Association
ITCA - International TeleConferencing Association
ITFS - Instructional Television Fixed Service
ITU-R - International Telecommunications Union- Radio sector (successor to CCIR)
ITU-T - International Telecommunications Union- Telecom sector (successor to CCITT)
IVCD - Initial Voice Channel Designation
IVR - Interactive Voice Response
IWF - InterWorking Function
IXC - Inter-eXchange Carrier
kHz - kiloHertz, 1000 cycles per second
Ki - secret identity key
KSU - Key Service Unit (small business telephone system)
LASTCHP - Last Channel Paging
LATA - Local Access and Transport Area
LEC - Local Exchange Carrier
LEO - Low Earth Orbit
LF - Low Frequency
LMCC - Land Mobile Communications Council
LMDS - Local Multipoint Distribution Service
LMR - Land Mobile Radio
MA - Mobile Allocation
MAC - Media Access Control
MACN - Mobile Allocation Channel Number
MAHO - Mobile Assisted Hand-Over (or Hand-Off)
MAI - Mobile Allocation Index
MAIO - Mobile Allocation Index Offset
MAP - Mobile Access Part (of CCS7, originally designed for use with GSM)
MCC - Mobile Country Code or Miscellaneous Common Carrier
MCI - Malicious Call Identification
MDS - Multipoint Distribution Service
MDT - Mobile Data Terminal
ME - Mobile Equipment
MEO - Medium Earth Orbit
MHz - MegaHertz, a million cycles per second
MIB - Management Information dataBase
MIN - Mobile Identification Number
MIRS - Motorola Integrated Radio System
MM - Man Machine
MMDS - Multichannel Multipoint Distribution Service
MMI - Man Machine Interface
MNC - Mobile Network Code

MNTA - Minnesota Telephone Association
MOC - Mobile-Originated Call
MODEM - MOdulator-DEModulator
ms - millisecond(s)
MS - Mobile Station
MSA - Metropolitan Statistical Area
MSC - Mobile service Switching Center
MSCM - Mobile Station Class Mark
MSIC - Mobile Subscriber Identification Number
MSISDN - Mobile Station ISDN Number
MSRN - Mobile Station Roaming Number
MSS - Mobile Satellite Service
MT - Message Type
MTA - Major Trading Area
MTC - Mobile Terminated Call
MTP - Message Transfer Protocol or Part
MTS - Mobile Telephone Service
MTSO - Mobile Telephone Switching Office
MTX - Mobile Telephone eXchange
NADC - North American Digital Cellular
N-AMPS - Narrowband AMPS
NAB - National Association of Broadcasters
NACN - North American Cellular Network
NAM - Number Assignment Module
NAMPS - Narrowband Advanced Mobile Phone Service
NANC - North American Numbering Council
NANP - North American Numbering Plan
NARUC - National Association of Regulatory Utility Commissioners
NASTD - National Association of State Telecommunications Directors
NAWC - Number of Additional Words Coming
N-CDMA - Narrow CDMA
NENA - National Emergency Number Association
NOI - Notice Of Inquiry
NPCS - Narrowband PCS, paging systems
NPRM - Notice of Proposed RuleMaking
NTACS - Narrowband TACS
NTCA - National Telephone Co-op Association
NTT-MCS - Nippon Telephone and Telegraph-Mobile Cellular System
NXX - end office code
OA&M - Operations, Administration and Maintenance
OCB - Outgoing Calls Barred
OLC - OverLoad Class
OPASTCO - Organization for the Promotion and Advancement of Small Telephone
COmpanies
PA - 1) Public Address (system), 2) Power Amplifier
PABX - Private Automatic Branch eXchange

PACS - Personal Access Communications System
PACT - Personal Air Communications Technology
PAGCH - Paging and Access Grant CHannel(s) (GSM related systems)
PBX - Private (automatic) Branch eXchange
PCCA - Portable Computer and Communications Association
PCH - Paging CHannel (GSM related systems)
PCI - Protocol Capability Indicator
PCIA - Personal Communications Industry Association
PCM - Pulse Cod(ed) Modulation
PCN - Personal Communication Network
PCS - Personal Communication Service
PCS-1900 - A modified version of the GSM system used on the 1900 MHz band
PDC - Pacific Digital Cellular
PDN - Public Data Networks
PHP - Personal HandyPhone
PHS - Personal Handyphone System
PIN - Personal Identification Number
PLL - Phase Locked Loop
PLMN - Public Land Mobile Network
PLMR - Public Land Mobile Radio
PLMS - Public Land Mobile Service
PM - Phase Modulation
PMR - Private Mobile Radio
PMRS - Private Mobile Radio Service
PMU - Personal Messaging Unit
PN - Pseudo-Noise
POCSAG - Post Office Code Standard Advisory Group
POP - Point Of Presence
POTS - Plain Old Telephone Service
PPDN - Public Packet Data Network
PRBS - Pseudonoise Random Binary String/Stream
PRI - Primary Rate Interface, an ISDN service connection at 1.544 or 2.048 Mb/s bit rate, with common channel packet signaling in the 24th channel of the first link
PROM - Programmable Read Only Memory
PSC - Public Service Commission
PSK - Phase Shift Keying
PSPDN - Packet Switched Public Data Network
PSTN - Public Switched Telephone Network
PTT - Postal Telephone and Telegraph or Push-To-Talk
PUC - Public Utilities Commission
QAM - Quadrature Amplitude Modulation
RA - Rural Area
RAB - Random Access Burst
RACE - Research and development of Advanced Communication technologies in Europe
RACH - Random Access CHannel
RAD - Remote Antenna Driver

RAM - Random Access Memory
RAO - Revenue Accounting Office
RASP - Remote Antenna Signal Processor
RCA - Rural Cellular Association
RCA - Radio Club of America
RCC - Radio Common Carrier
RDTC - Reverse Digital Traffic Channel
RECC - REverse analog Control Channel
RELP - Residual (or regular pulse) Excited Linear Predictive (speech coder)
RF - Radio Frequency
RFC - Radio Frequency Channel
ROM - Read Only Memory
RP - Radio Port
RPCU - Radio Port Control Unit
RPE - Regular Pulse Excitation
RR - Radio Resource
RSA - Reliable Service Area (for paging) or Rural Service Area (for cellular)
RSA - Reliable Service Area
RSL - Radio Signaling Link
RSSI - Received Signal Strength Indicator (or Indication)
RTC - Reverse Traffic Channel digital
RVC - Reverse analog Voice Channel
RX - (radio) receiver
S/I - Signal-to-Interference ratio
S/N - Signal-to-Noise ratio
S7 - Signaling system 7, alternate abbreviation for CCS7
SAT - Supervisory Audio Tone (analog cellular systems)
SBCA - Satellite Broadcasting and Communications Association
SC - Signaling Controller
SC - Service Center
SCCP - Signaling Connection Control Part
SCCS - Selective Class of Call Screening
SCM - Station Class Mark
SCP - Service Control Point
SCTE - Society of Cable and Television Engineers
SDCCH - Stand alone Dedicated Control Channel
SDMA - Spacial Division Multiple Access
SFH - Slow Frequency Hopping
SHF - Super High Frequency
SIA - Satellite Industry Association
SID - System IDentification
SIM - Subscriber Identity Module
SIMEG - Subscriber Identity Module Expert Group
SINAD - SIgnal+Noise And Distortion
SMG - Special Mode Group
SMR - Specialized Mobile Radio (portion of 800 MHz band in North America)

SMS - Service Management System or Short Messaging Service
SMSCB - Short Message Service Call Broadcast
SMT - Surface Mounted Technology
SNR - Signal-to-Noise Ratio
SONET - Synchronous Optical NETwork
SP - Signaling Point
SRES - System RESponse (related to authentication)
SS - Supplementary Service(s)
SS7 - alternate abbreviation for CCS7 (signaling system 7)
SSB - Single Side Band, a special form of AM
SSD - Shared Secret Data
SSN - Signaling System Number
SSP - Service Switching Point
ST - Signaling Tone
SWR - Standing Wave Ratio
T/R - Transmit or Receive
TACH - Traffic and Associated CHannel(s) (GSM and related systems)
TACS - Total Access Communications System
TASI - Time Assignment Speech Interpolation (see DSI)
TCH - Traffic CHannel
TCH/FS - Traffic CHannel/Full rate Speech
TCH/HS - Traffic CHannel/Half rate Speech
TCM - Time Compression Multiplexing
TCP/IP - Transmission Control Protocol/Internet Protocol
TDD - Time Division Duplex
TDM - Time Division Multiplexing
TDMA - Time Division Multiple Access
TEI - Terminal Equipment Identifier
TIA - Telecommunications Industry Association
TMN - Telecommunication Management Network
TMSI - Temporary Mobile Service Identity (GSM and related systems)
TRA - Telecommunications Resellers Association
Tx - transmitter, transmit
uCell - micro Cell (u is used here as a substitute for Greek letter Mu, written μ)
UCH - User CHannel
UHF - Ultra-High Frequency, specifically frequencies between 300 MHz and 3 GHz
UI - Unnumbered Information
Um - radio link interface between MS and BS
UMTS - Universal Mobile Telecommunications System
uP - microProcessor (u is used here as a substitute for Greek letter Mu, written μ)
UPR - User Performance Requirements
UPT - Universal Personal Telecommunications
uSec - microSecond
USTA - United States Telephone Association
UTC - United Telecommunications Council
UUT - Unit Under Test

UWCC - United Wireless Communications Consortium
VAD - Voice Activity Detection
VCO - Voltage Controlled Oscillator
VCS - Voice Controlled Switch(ing)
VHF - Very High Frequency
VLF - Very Low Frequency
VLR - Visited Location Register
VMAC - Voice Mobile Attenuation Code
VOX - Voice control(led)
VSWR - Voltage Standing Wave Ratio
WACP - Wide Area Calling Plan
WACS - Wireless Access Communications System
WARC - World Administration of Radio Conference
WATS - Wide Area Telephone Service
WB-CDMA - WideBand Code Division Multiple Access
WC - Wireless Carrier
WCA - Wireless Cable Association international
WDA - Wireless Dealers Association
WDF - Wireless Data Forum
WER - Word Error Rate
WFOM - Wait For Overhead Message
WIA - Wireless Industry Association
WLA - Wireless LAN Alliance
WLL - Wireless Local Loop
WOTS - Wireless Office Telephone System
WPBX - Wireless PBX
WSC - Wireless Switching Center
WSP - Wireless Service Provider
X.25 - ITU standard describing a packet data network, and an implementation of that standard
in a working network
ZBTSI - Zero Byte Time Slot Interchange, a form of line coding which permits consecutive
groups of 8 binary zeros on digital transmission links in the PSTN

Appendix II - Industry Standards

Document Distribution

Standard documents can be obtained from Global Engineering Documents (GED) at the offices listed below.

United States

Global Engineering Documents
15 Inverness Way East
Englewood, COLORADO, 80112 USA
E-Mail: global@ihs.com
Phone: +1-800-854-7179
Phone: +1-303-397-7956 (outside the U.S.)
Fax: +1-303-397-2740

Australia

Global Info Centre
IHS Australia Pty. Ltd.
Building B, Level 3
242 Beecroft Road, Epping NSW 2121 AUSTRALIA
E-Mail: global.australia@ihs.com
Phone: +61+2 9876 5333
Fax: +61+2 9876 2299

Brazil

Global Info Centre do Brasil, Ltda.
Rua Clodomiro Amazonas
1099 - cj. - 105
Sao Paulo - SP - BRASIL - CEP 04537-012
E-Mail: global.brazil@ihs.com
Phone: (011)866-7194
Fax: (011)866-7193

Canada

Global Info Centre
IHS Canada/Micromedia
240 Catherine Street, Suite 305
Ottawa, Ontario K2P 2G8 CANADA
E-Mail: gic@ihs.on.ca
Phone: 613-237-4250
Fax: 613-237-4251

Colombia

Global Info Centre
Infoenlace Ltd.
Carrera 6 No. 51-21 A.A. 34270
Bogota COLOMBIA
E-Mail: infoenlace@gaitana.interred.net.co
Phone: +57-1-285-1779
Fax: +57-1-285-2798

France

Global Info Centre
31-35 rue deNeuilly
92110 Clichy FRANCE
E-Mail: global.paris@ihs.com
Phone: +33-1-40-87-17-02
Fax: +33-1-40-87-07-52

Germany

INFORMATION HANDLING SERVICES GmbH
Global Info Centre
Fraunhofer Str. 22
D-82152 Planegg (Martinsried) GERMANY
E-Mail: mail@ihs.de
Phone: +49-89-8952690
Fax: +49-89-89526999

Hong Kong

> Global Info Centre
> IHS Hong Kong Ltd.
> Unit 1,11F., Multifield Plaza, 3-7A Prat Avenue
> Tsim Sha Tsui, Kowloon HONG KONG
> E-Mail: global.hongkong@ihs.com
> Phone: +852-2368-5733
> Fax: +852-2368-5269

Italy

> Global Info Centre Rome
> Via Del Tritone 132
> 00187 ROME (RM) ITALY
> E-Mail: global.italy@pronet.it
> Phone: +39-06-4786833
> Fax: +39-06-4821813

Mexico

> Global Info Centre
> IHS de Mexico S.A. de C.V.
> Dante #36 Piso 10
> Col. Anzures Mexico, D.F. C.P. 11590
> E-Mail: global.mexico@ihs.com
> Phone: +52-5-250-0611
> Fax: +52-5-254-5815

Latin America (via Miami)

> Global Engineering Documents
> Global Info Centres
> 6303 Blue Lagoon Drive, Suite 370
> Miami, FL 33126 USA
> E-Mail: global.csa@ihs.com
> Phone: 305-262-4022
> Fax: 305-262-8232

Nordic

Global Info Centre
IHS Nordic A.S.
Strandvejen 130
DK - 2900 Hellerup DENMARK
E-Mail: info@ihsnordic.com
Phone: +45 39 40 22 44
Fax: +45 39 40 23 00

South Africa

Global Info Centre - South Africa
Corner Northern Pkwy/Handel Road
Ormonde Extension 5
Johannesburg 2091 South Africa
E-Mail: global@ihs.co.za
Phone: +27-11-835-2221
Fax: +27-11-835-1943

United Kingdom

RAPIDOC(R)
Willoughby Road
Bracknell
Berkshire RGl2 8DW UNITED KINGDOM
E-Mail: rapidoc@techindex.co.uk
Phone: +44-1344-861666
Fax: +44-1344-714440

TERRESTRIAL TRUNKED RADIO (TETRA)

ETS 300 392-10-03
        TERRESTRIAL TRUNKED RADIO
        (TETRA); VOICE PLUS DATA (V+D);
        PART 10: SUPPLEMENTARY
        SERVICES STAGE 1; SUB-PART 3:
        TALKING PARTY IDENTIFICATION (TPI)

ETS 300 392-10-1
    TERRESTRIAL TRUNKED RADIO
    (TETRA); VOICE PLUS DATA (V+D);
    PART 10: SUPPLEMENTARY
    SERVICES STAGE 1; SUB-PART 1:
    CALL IDENTIFICATION

ETS 300 392-10-1
DRAFT
    TERRESTRIAL TRUNKED RADIO
    (TETRA); VOICE PLUS DATA (V+D);
    PART 10: SUPPLEMENTARY
    SERVICES STAGE 1; SUB-PART 1:
    CALL INDENTIFICATION

ETS 300 392-10-13
DRAFT
    TERRESTRIAL TRUNKED RADIO
    (TETRA); VOICE PLUS DATA (V+D);
    PART 10: SUPPLEMENTARY
    SERVICES STAGE 1; SUB-PART 13:
    CALL COMPLETION TO BUSY
    SUBSCRIBER

ETS 300 392-10-20
    TERRESTRIAL TRUNKED RADIO
    (TETRA); VOICE PLUS DATA (V+D);
    PART 10: SUPPLEMENTARY
    SERVICES STAGE 1; SUB-PART 20:
    DISCREET LISTENING (DL)

ETS 300 392-10-20
DRAFT
    TERRESTRIAL TRUNKED RADIO
    (TETRA); VOICE PLUS DATA (V+D);
    PART 10: SUPPLEMENTARY
    SERVICES STAGE 1; SUB-PART 20:
    DISCREET LISTENING (DL)

ETS 300 392-10-23
DRAFT
    TERRESTRIAL TRUNKED RADIO
    (TETRA); VOICE PLUS DATA (V+D);
    PART 10: SUPPLEMENTARY
    SERVICES STAGE 1; SUB-PART 23:
    CALL COMPLETION ON NO REPLY

ETS 300 392-10-3
DRAFT
    TERRESTRIAL TRUNKED RADIO
    (TETRA); VOICE PLUS DATA (V+D);
    PART 10: SUPPLEMENTARY
    SERVICES STAGE 1; SUB-PART 3:
    TALKING PARTY IDENTIFICATION (TPI)

ETS 300 392-10-6
    TERRESTRIAL TRUNKED RADIO
    (TETRA); VOICE PLUS DATA (V+D);
    PART 10: SUPPLEMENTARY
    SERVICES STAGE 1; SUB-PART 6:
    CALL AUTHORIZED BY DISPATCHER
    (CAD)

ETS 300 392-10-7
DRAFT
    TERRESTRIAL TRUNKED RADIO
    (TETRA); VOICE PLUS DATA (V+D);
    PART 10: SUPPLEMENTRY SERVICES
    STAGE 1; SUB-PART 7: SHORT
    NUMBER ADDRESSING

ETS 300 392-10-9
    TERRESTRIAL TRUNKED RADIO
    (TETRA); VOICE PLUS DATA (V+D);
    PART 10: SUPPLEMENTARY
    SERVICES STAGE 1; SUB-PART 9:
    ACCESS PRIORITY

ETS 300 392-11-03
    TERRESTRIAL TRUNKED RADIO
    (TETRA); VOICE PLU DATA (V+D); PART
    11: SUPPLEMENTARY SERVICES
    STAGE 2; SUB-PART 3: TALKING
    PARTY IDENTIFICATION (TPI)

ETS 300 392-11-1

> TERRESTRIAL TRUNKED RADIO
> (TETRA); VOICE PLUS DATA (V+D);
> PART 11: SUPPLEMENTARY
> SERVICES STAGE; SUB-PART 1: CALL
> IDENTIFICATION (CI)

ETS 300 392-11-1
DRAFT

> TERRESTRIAL TRUNKED RADIO
> (TETRA); VOICE PLUS DATA (V+D);
> PART 11: SUPPLEMENTARY
> SERVICES STAGE; SUB-PART 1: CALL
> IDENTIFICATION (CI)

ETS 300 392-11-13

> TERRESTRIAL TRUNKED RADIO
> (TETRA); VOICE PLUS DATA (V+D);
> PART 11: SUPPLEMENTARY SERVICES
> STAGE 2; SUB-PART 13: CALL
> COMPLETION TO BUSY SUBSCRIBER
> (CCBS)

ETS 300 392-11-13
DRAFT

> TERRESTRIAL TRUNKED RADIO
> (TETRA); VOICE PLUS DATA (V+D);
> PART 11: SUPPLEMENTARY
> SERVICES STAGE 2; SUB-PART 13:
> CALL COMPLETION TO BUSY
> SUBSCRIBER (CCBS)

ETS 300 392-11-20

> TERRESTRIAL TRUNKED RADIO
> (TETRA); VOICE PLUS DATA (V+D);
> PART 11: SUPPLEMENTARY
> SERVICES STAGE 2; SUB-PART 20:
> DISCREET LISTENING (DL)

Public and Private Land Mobile Radio Telephones and Systems

ETS 300 392-11-20
DRAFT
>   TERRESTRIAL TRUNKED RADIO
>   (TETRA); VOICE PLUS DATA (V+D);
>   PART 11: SUPPLEMENTARY
>   SERVICES STAGE 2; SUB-PART 20:
>   DISCREET LISTENING (DL)

>   TERRESTRIAL TRUNKED RADIO
>   (TETRA); VOICE PLUS DATA (V+D);
>   PART 11: SUPPLEMENTARY
>   SERVICES STAGE 2; SUB-PART 23:
>   CALL COMPLETION ON NO REPLY
>   (CCNR)

ETS 300 392-11-23
DRAFT
>   TERRESTRIAL TRUNKED RADIO
>   (TETRA); VOICE PLUS DATA (V+D);
>   PART 11: SUPPLEMENTARY
>   SERVICES STAGE 2; SUB-PART 23:
>   CALL COMPLETION ON NO REPLY
>   (CCNR)

ETS 300 392-11-3
>   TERRESTRIAL TRUNKED RADIO
>   (TETRA); VOICE PLUS DATA (V+D);
>   PART 11 SUPPLEMENTARY SERVICES
>   STAGE 2: SUB-PART 3: TALKING
>   PARTY IDENTIFICATION (TPI)

ETS 300 392-11-3
DRAFT
>   TERRESTRIAL TRUNKED RADIO
>   (TETRA); VOICE PLUS DATA (V+D);
>   PART 11: SUPPLEMENTARY
>   SERVICES STAGE 2: SUB-PART 3:
>   TALKING PARTY IDENTIFICATION (TPI)

ETS 300 392-11-6

> TERRESTRIAL TRUNKED RADIO
> (TETRA); VOICE PLUS DATA (V+D);
> PART 11: SUPPLEMENTARY
> SERVICES STAGE 2; SUB-PART 6:
> CALL AUTHORIZED BY DISPATCHER
> (CAD)

ETS 300 392-11-7

> TERRESTRIAL TRUNKED RADIO
> (TETRA); VOICE PLUS DATA (V+D)
> PART 11: SUPPLEMENTARY
> SERVICES STAGE 2; SUB-PART 7:
> SHORT NUMBER ADDRESSING (SNA)

ETS 300 392-11-7
DRAFT

> TERRESTRIAL TRUNKED RADIO
> (TETRA); VOICE PLUS DATA (V+D);
> PART 11: SUPPLEMENTARY
> SERVICES STAGE 2; SUB-PART 7:
> SHORT NUMBER ADDRESSING (SNA)

ETS 300 392-11-9

> TERRESTRIAL TRUNKED RADIO
> (TETRA); VOICE PLUS DATA (V+D);
> PART 11: SUPPLEMENTARY SERVICES
> STAGE 2; SUB-PART 9: ACCESS
> PRIORITY (AP)

ETS 300 392-12-03

> TERRESTRIAL TRUNKED RADIO
> (TETRA); VOICE PLUS DATA (V+D);
> PART 12: SUPPLEMENTARY SERVICES
> STAGE 3; SUB-PART 3: TALKING
> PARTY IDENTIFICATION (TPI)

ETS 300 392-12-1

> TERRESTRIAL TRUNKED RADIO
> (TETRA); VOICE PLUS DATA (V+D);
> PART 12: SUPPLEMENTARY SERVICES
> STAGE 3; SUB-PART 1: CALL
> IDENTIFICATION (CI)

ETS 300 392-12-1
DRAFT
    TERRESTRIAL TRUNKED RADIO
    (TETRA); VOICE PLUS DATA (V+D);
    PART 12: SUPPLEMENTARY SERVICES
    STAGE 3; SUB-PART 1: CALL
    IDENTIFICATION (CI)

ETS 300 392-12-13
    TERRESTRIAL TRUNKED RADIO
    (TETRA) VOICE PLUS DATA (V+D);
    PART 12: SUPPLEMENTARY SERVICES
    STAGE 3; SUB-PART 13: CALL
    COMPLETION TO BUSY SUBSCRIBER
    (CCBS)

ETS 300 392-12-13
DRAFT
    TERRESTRIAL TRUNKED RADIO
    (TETRA); VOICE PLUS DATA (V+D);
    PART 12: SUPPLEMENTARY SERVICES
    STAGE 3; SUB-PART 13: CALL
    COMPLETION TO BUSY SUBSCRIBER
    (CCBS)

ETS 300 392-12-20
    TERRESTRIAL TRUNKED RADIO
    (TETRA); VOICE PLUS DATA (V+D);
    PART 12: SUPPLEMENTARY SERVICES
    STAGE 3 DESCRIPTION; SUB-PART 20:
    DISCREET LISTENING (DL)

ETS 300 392-12-20
DRAFT
    TERRESTRIAL TRUNKED RADIO
    (TETRA); VOICE PLUS DATA (V+D);
    PART 12: SUPPLEMENTARY SERVICES
    STAGE 3 DESCRIPTION; SUB-PART 20:
    DISCREET LISTENING (DL)

ETS 300 392-12-23
>TERRESTRIAL TRUNKED RADIO
(TETRA); VOICE PLUS DATA (V+D);
PART 12: SUPPLEMENTARY SERVICES
STAGE 3; SUB-PART 23: CALL
COMPLETION ON NO REPLY (CCNR)

ETS 300 392-12-23
DRAFT
>TERRESTRIAL TRUNKED RADIO
(TETRA); VOICE PLUS DATA (V+D);
PART 12: SUPPLEMENTARY SERVICES
STAGE 3; SUB-PART 23: CALL
COMPLETION ON NO REPLY (CCNR)

ETS 300 392-12-3
>TERRESTRIAL TRUNKED RADIO
(TETRA); VOICE PLUS DATA (V+D);
PART 12: SUPPLEMENTARY SERVICES
STAGE 3; SUB-PART TALKING PARTY
IDENTIFICATION (TPI)

ETS 300 392-12-3
DRAFT
>TERRESTRIAL TRUNKED RADIO
(TETRA); VOICE PLUS DATA (V+D);
PART 12: SUPPLEMENTARY SERVICES
STAGE 3; SUB-PART TALKING PARTY
IDENTIFICATION (TPI)

ETS 300 392-12-6
>TERRESTRIAL TRUNKED RADIO
(TETRA); VOICE PLUS DATA (V+D);
PART 12: SUPPLEMENTARY SERVICES
STAGE 3; SUB-PART 6: CALL
AUTHORIZED BY DISPATCHER (CAD)

ETS 300 392-12-7
>TERRESTRIAL TRUNKED RADIO
(TETRA); VOICE PLUS DATA (V+D);
PART 12: SUPPLEMENTARY SERVICES
STAGE 3; SUB-PART 7: SHORT
NUMBER ADDRESSING (SNA)

ETS 300 392-12-7
DRAFT
    TERRESTRIAL TRUNKED RADIO
    (TETRA) VOICE PLUS DATA (V+D);
    PART 12; SUPPLEMENTARY SERVICES
    STAGE 3; SUB-PART 7: SHORT
    NUMBER ADDRESSING (SNA)

ETS 300 392-12-9
    TERRESTRIAL TRUNKED RADIO
    (TETRA); VOICE PLUS DATA
    (V+D); PART 12: SUPPLEMENTARY
    SERVICES STAGE 3; SUB-PART 9:
    ACCESS PRIORITY (AP)

ETS 300 392-3-2
    TERRESTRIAL TRUNKED RADIO
    (TETRA); VOICE PLUS DATA (V+D);
    PART 3: INTERWORKING AT THE
    INTER-SYSTEM INTERFACE (ISI);
    SUB-PART 2: ADDITIONAL NETWORK
    FUNCTIONS INDIVIDUAL CALL
    (ANF-ISIIC)

ETS 300 392-3-2
    TERRESTRIAL TRUNKED RADIO
    (TETRA); VOICE PLUS DATA (V+D);
    PART 3: INTERWORKING AT THE
    INTER-SYSTEM INTERFACE (ISI);
    SUB-PART 2: ADDITIONAL NETWORK
    FUNCTIONS INDIVIDUAL CALL
    (ANF-ISIIC)

ETS 300 392-3-3
    TERRESTRIAL TRUNKED RADIO
    (TETRA); VOICE PLUS DATA (V+D);
    PART 3: INTERWORKING AT THE
    INTER-SYSTEM INTERFACE (ISI);
    SUB-PART 3: ADDITIONAL NETWORK
    FUNCTIONS GROUP CALL (ANF-ISIGC)

ETS 300 392-3-3
DRAFT
      TERRESTRIAL TRUNKED RADIO
      (TETRA); VOICE PLUS DATA (V+D)
      PART 3: INTERWORKING AT THE
      INTER-SYSTEM INTERFACE (ISI);
      SUB-PART 3: ADDITIONAL NETWORK
      FUNCTIONS GROUP CALL (ANF-ISIGC)

ETS 300 392-3-5
      TERRESTRIAL TRUNKED RADIO
      (TETRA); VOICE PLUS DATA (V+D);
      PART 3: INTERWORKING AT
      INTER-SYSTEM INTERFACE (ISI);
      SUB-PART 5: ADDITIONAL NETWORK
      FEATURE FOR MOBILITY
      MANAGEMENT (ANF-ISIMM)

ETS 300 392-3-5
      TERRESTRIAL TRUNKED RADIO
      (TETRA); VOICE PLUS DATA (V+D);
      PART 3: INTERWORKING AT
      INTER-SYSTEM INTERFACE (ISI);
      SUB-PART 5: ADDITIONAL NETWORK
      FEATURE FOR MOBILITY
      MANAGEMENT (ANF-ISIMM)

ETS 300 392-4-2
      TERRESTRIAL TRUNKED RADIO
      (TETRA); VOICE PLUS DATA (V+D)
      PART 4: GATEWAY BASIC OPERATION;
      SUB-PART 2: INTEGRATED SERVICE
      DIGITAL NETWORK (ISDN) GATEWAY

ETS 300 392-4-2
DRAFT
      TERRESTRIAL TRUNKED RADIO
      (TETRA); VOICE PLUS DATA (V+D)
      PART 4: GATEWAY BASIC OPERATION;
      SUB-PART 2: INTEGRATED SERVICE
      DIGITAL NETWORK (ISDN) GATEWAY

ETS 300 392-4-3
>TERRESTRIAL TRUNKED RADIO
>(TETRA); VOICE PLUS DATA (V+D);
>PART 4: GATEWAYS BASIC
>OPERATION; SUB-PART 3: DATA
>NETWORKS GATEWAY

ETS 300 392-4-3
DRAFT
>TERRESTRIAL TRUNKED RADIO
>(TETRA); VOICE PLUS DATA (V+D);
>PART 4: GATEWAYS BASIC
>OPERATION; SUB-PART 3: DATA
>NETWORKS GATEWAY

ETS 300 392-7
DRAFT
>TERRESTRIAL TRUNKED RADIO
>(TETRA); VOICE PLUS DATA (V+D); PART 7:
>SECURITY

ETS 300 393-10
>TERRESTRIAL TRUNKED RADIO
>(TETRA); PACKET DATA OPTIMIZED
>(PDO); PART 10: SDL MODEL OF AIR
>INTERFACE (AI)

ETS 300 393-10
DRAFT
>TERRESTRIAL TRUNKED RADIO
>(TETRA); PACKET DATA OPTIMIZED
>(PDO); PART 10: SDL MODEL OF AIR
>INTERFACE (AI)

ETS 300 393-11
>TERRESTRIAL TRUNKED RADIO
>(TETRA); PACKET DATA OPTIMIZED
>(PDO); PART 11: PROTOCOL
>IMPLEMENTATION CONFORMANCE
>STATEMENT (PICS) PROFORMA
>SPECIFICATION

ETS 300 393-2
  TERRESTRIAL TRUNKED RADIO
  (TETRA); PACKET DATA OPTIMIZED
  (PDO); PART 2: AIR INTERFACE (AI)

ETS 300 393-2
DRAFT
  TERRESTRIAL TRUNKED RADIO
  (TETRA); PACKET DATA OPTIMIZED
  (PDO); PART 2: AIR INTERFACE (AI)

ETS 300 394-4-1
  TERRESTRIAL TRUNKED RADIO
  (TETRA); CONFORMANCE TESTING
  SPECIFICATION; PART 4: PROTOCL
  TESTING SPECIFICATION FOR DIRECT
  MODE OPERATION (DMO); SUB-PART
  1: TEST SUITE STRUCTURE AND TEST
  PURPOSES (TSS&TP) FOR MOBILE
  STATION TO MOBILE STATION (MS-MS)
  AIR

ETS 300 394-4-1
DRAFT
  TERRESTRIAL TRUNKED RADIO
  (TETRA); CONFORMANCE TESTING
  SPECIFICATION; PART 4: PROTOCOL
  TESTING SPECIFICATION FOR DIRECT
  MODE OPERATION (DMO); SUB-PART
  1: TEST SUITE STRUCTURE AND TEST
  PURPOSES (TSS&TP) FOR MOBILE
  STATION TO MOBILE STATION (MS-MS)
  AIR

ETS 300 394-4-10
  TERRESTRIAL TRUNKED RADIO
  (TETRA); CONFORMANCE TESTING
  SPECIFICATION; PART 4: PROTOCOL
  TESTING SPECIFICATION FOR DIRECT
  MODE OPERATION (DMO); SUB-PART
  10: ABSTRACT TEST SUITE (ATS) FOR
  DIRECT MODE GATEWAY (DM-GATE)

ETS 300 394-4-10
DRAFT
  TERRESTRIAL TRUNKED RADIO
  (TETRA); CONFORMANCE TESTING
  SPECIFICATION; PART 4: PROTOCOL
  TESTING SPECIFICATION FOR DIRECT
  MODE OPERATION (DMO); SUB-PART
  10: ABSTRACT TEST SUITE (ATS) FOR
  DIRECT MODE GATEWAY (DM-GATE)

ETS 300 394-4-2
  TERRESTRIAL TRUNKED RADIO
  (TETRA); CONFORMANCE TESTING
  SPECIFICATION; PART 4: PROTOCOL
  TESTING SPECIFICATION FOR DIRECT
  MODE OPERATION (DMO); SUB-PART
  2: ABSTRACT TEST SUITES (ATS) FOR
  MOBILE STATION TO MOBILE STATION
  (MS-MS) AIR INTERFACE (AI)

ETS 300 394-4-2
DRAFT
  TERRESTRIAL TRUNKED RADIO
  (TETRA); CONFORMANCE TESTING
  SPECIFICATION; PART 4: PROTOCOL
  TESTING SPECIFICATION FOR DIRECT
  MODE OPERATION (DMO); SUB-PART
  2: ABSTRACT TEST SUITES (ATS) FOR
  MOBILE STATION TO MOBILE STATION
  (MS-MS) AIR INTERFACE (AI)

ETS 300 394-4-7
  TERRESTRIAL TRUNKED RADIO
  (TETRA); PART 4: PROTOCOL TESTING
  SPECIFICATION FOR DIRECT MODE
  OPERATION (DMO); SUB-PART 7: TEST SUITE
  STRUCTURE AND TEST PURPOSES
  (TSS&TP) FOR MOBILE STATION TO
  GATEWAY (MS-GW) AIR INTERFACE
  (AI)

ETS 300 394-4-7
DRAFT

>TERRESTRIAL TRUNKED RADIO
>(TETRA); PART 4: PROTOCOL TESTING
>SPECIFICATION FOR DIRECT MODE
>OPERATION (DMO); SUB-PART 7: TEST SUITE
>STRUCTURE AND TEST PURPOSES
>(TSS&TP) FOR MOBILE STATION TO
>GATEWAY (MS-GW) AIR INTERFACE
>(AI)

ETS 300 394-4-8

>TERRESTRIAL TRUNKED RADIO
>(TETRA); CONFORMANCE TESTING
>SPECIFICATION; PART 4: PROTOCOL
>TESTING SPECIFICATION FOR DIRECT
>MODE OPERATION (DMO); SUB-PART
>8: TEST SUITE STRUCTURE AND TEST
>PURPOSES (TSS&TP) FOR DIRECT
>MODE GATEWAY (DM-GATE)

ETS 300 394-4-8
DRAFT

>TERRESTRIAL TRUNKED RADIO
>(TETRA); CONFORMANCE TESTING
>SPECIFICATION; PART 4: PROTOCOL
>TESTING SPECIFICATION FOR DIRECT
>MODE OPERATION (DMO); SUB-PART
>8: TEST SUITE STRUCTURE AND TEST
>PURPOSES (TSS&TP) FOR DIRECT
>MODE GATEWAY (DM-GATE)

ETS 300 394-4-9

>TERRESTRIAL TRUNKED RADIO
>(TETRA); CONFORMANCE TESTING
>SPECIFICATION; PART 4: PROTOCOL
>TESTING SPECIFICATION FOR DIRECT
>MODE OPERATION (DMO); SUB-PART 9:
>ABSTRACT TEST SUITE (ATS) FOR
>MOBILE STATION (MS) GATEWAY

ETS 300 394-4-9
DRAFT
> TERRESTRIAL TRUNKED RADIO
> (TETRA); CONFORMANCE TESTING
> SPECIFICATION; PART 4: PROTOCOL
> TESTING SPECIFICATION FOR DIRECT
> MODE OPERATION (DMO); SUB-PART
> 9: ABSTRACT TEST SUITE (ATS) FOR
> MOBILE STATION (MS) GATEWAY

ETS 300 394-5-1
> TERRESTRIAL TRUNKED RADIO
> (TETRA); CONFORMANCE TESTING
> SPECIFICATION; PART 5: SECURITY;
> SUB-PART 1: PROTOCOL
> IMPLEMENTATION CONFORMANCE
> STATEMENT (PICS) PROFORMA
> SPECIFICATION

ETS 300 394-5-1
DRAFT
> TERRESTRIAL TRUNKED RADIO
> (TETRA); CONFORMANCE TESTING
> SPECIFICATION; PART 5: SECURITY;
> SUB-PART 1: PROTOCOL
> IMPLMENTATION CONFORMANCE
> STATEMENT (PICS) PROFORMA
> SPECIFICATION

ETS 300 394-5-2
> TERRESTRIAL TRUNKED RADIO
> (TETRA); CONFORMANCE TESTING
> SPECIFICATION; PART 5; SECURITY;
> SUB-PART 2: PROTOCOL TESTING
> SPECIFICATION FOR TETRA SECURITY

ETS 300 394-5-2
DRAFT
> TERRESTRIAL TRUNKED RADIO
> (TETRA); CONFORMANCE TESTING
> SPECIFICATION; PART 5: SECURITY
> SUB-PART 2: PROTOCOL TESTING
> SPECIFICATION FOR TETRA SECURITY

ETS 300 394-5-3

> TERRESTRIAL TRUNKED RADIO
> (TETRA); CONFORMANCE TESTING
> SPECIFICATION; PART 5: SECURITY;
> SUB-PART 3: ABSTRACT TEST SUITE
> (ATS)

ETS 300 394-5-3
DRAFT

> TERRESTRIAL TRUNKED RADIO
> (TETRA); CONFORMANCE TESTING
> SPECIFICATION; PART 5: SECURITY;
> SUB-PART 3: ABSTRACT TEST SUITE
> (ATS)

ETS 300 396-7

> TERRESTRIAL TRUNKED RADIO
> (TETRA); TECHNICAL REQUIREMENTS
> FOR DIRECT MODE OPERATION (DMO);
> PART 7: REPEATER TYPE 2

ETS 300 396-7
DRAFT

> TERRESTRIAL TRUNKED RADIO
> (TETRA); TECHNICAL REQUIREMENTS
> FOR DIRECT MODE OPERATION (DMO);
> PART 7: REPEATER TYPE 2

ETS 300 396-8-3

> TERRESTRIAL TRUNKED RADIO
> (TETRA); TECHNICAL REQUIREMENTS
> FOR DIRECT MODE OPERATION (DMO);
> PART 8: PROTOCOL IMPLEMENTATION
> CONFORMANCE STATEMENT (PICS)
> PROFORMA SPECIFICATION; SUB-PART
> 3: GATEWAY AIR INTERFACE (AI)

ETS 300 396-8-3
DRAFT
     TERRESTRIAL TRUNKED RADIO
     (TETRA); TECHNICAL REQUIREMENTS
     FOR DIRECT MODE OPERATION (DMO);
     PART 8: PROTOCOL IMPLEMENTATION
     CONFORMANCE STATEMENT (PICS)
     PROFORMA SPECIFICATION; SUB-PART
     3: GATEWAY AIR INTERFACE (AI)

ETS 300 812
     TERRESTRIAL TRUNKED RADIO
     (TETRA); SECURITY ASPECTS;
     SUBSCRIBER IDENTITY MODULE TO
     MOBILE EQUIPMENT (SIM - ME)
     INTERFACE

ETS 300 812
     TERRESTRIAL TRUNKED RADIO
     (TETRA); SECURITY ASPECTS;
     SUBSCRIBER IDENTITY MODULE TO
     MOBILE EQUIPMENT (SIM - ME)
     INTERFACE

ETSI EG 201 040
     TERRESTRIAL TRUNKED RADIO
     (TETRA); SECURITY; LAWFUL
     INTERCEPTION (LI) INTERFACE;
     FEASIBILITY STUDY REPORT

ETSI EN 300 827
     ELECTROMAGNETIC COMPATIBILITY
     AND RADIO SPECTRUM MATTERS
     (ERM); ELECTROMAGNETIC
     COMPATIBILITY (ERC) STANDARD FOR
     TERRESTRIAL TRUNKED RADIO (TETRA)
     AND ANCILLARY EQUIPMENT

ETSI EN 300 827

> ELECTROMAGNETIC COMPATIBILITY
> AND RADIO SPECTRUM MATTERS
> (ERM); ELECTROMAGNETIC
> COMPATIBILITY (ERC) STANDARD FOR
> TERRESTRIAL TRUNKED RADIO (TETRA)
> AND ANCILLARY EQUIPMENT
>
> TERRESTRIAL TRUNKED RADIO
> (TETRA); SECURITY; LAWFUL
> INTERCEPTION (LI) INTERFACE

ETSI EN 301 747

> TERRESTRIAL TRUNKED RADIO
> (TETRA); VOICE PLUS DATA (V+D); IP
> INTERWORKING (IPI)

ETSI ES 201 658

> TERRESTRIAL TRUNKED RADIO
> (TETRA); DIGITAL ADVANCED
> WIRELESS SERVICE (DAWS); LOGICAL
> LINK CONTROL (LLC) SERVICE
> DESCRIPTION

ETSI ES 201 659

> TERRESTRIAL TRUNKED RADIO
> (TETRA); DIGITAL ADVANCED
> WIRELESS SERVICE (DAWS); MEDIUM
> ACCESS CONTROL (MAC) SERVICE
> DESCRIPTION

ETSI ES 201 659
DRAFT

> TERRESTRIAL TRUNKED RADIO
> (TETRA); DIGITAL ADVANCED
> WIRELESS SERVICE (DAWS); MEDIUM
> ACCESS CONTROL (MAC) SERVICE
> DESCRIPTION

ETSI ES 201 660

> TERRESTRIAL TRUNKED RADIO
> (TETRA); DIGITAL ADVANCED
> WIRELESS SERVICE (DAWS);
> PHYSICAL LAYER (PHY) SERVICE
> DESCRIPTION

ETSI ES 201 660
DRAFT
    TERRESTRIAL TRUNKED RADIO
    (TETRA); DIGITAL ADVANCED
    WIRELESS SERVICE (DAWS);
    PHYSICAL LAYER (PHY) SERVICE
    DESCRIPTION

ETSI ETR 292
    TERRESTRIAL TRUNKED RADIO
    (TETRA); VOICE PLUS DATA (V+D);
    TECHNICAL REQUIREMENTS
    SPECIFICATION; NETWORK
    MANAGEMENT

ETSI ETR 300-1
    TERRESTRIAL TRUNKED RADIO
    (TETRA); VOICE PLUS DATA (V+D);
    DESIGNERS' GUIDE; PART 1:
    OVERVIEW, TECHNICAL DESCRIPTION
    AND RADIO ASPECTS

ETSI ETR 300-2
    TERRESTRIAL TRUNKED RADIO
    (TETRA); VOICE PLUS DATA (V+D);
    DESIGNERS GUIDE: PART 2 RADIO
    CHANNELS, NETWORK PROTOCOL AND
    SERVICES PERFORMANCE

ETSI ETR 300-4
    TERRESTRIAL TRUNKED RADIO
    (TETRA); VOICE PLUS DATA (V+D);
    DESIGNERS' GUIDE; PART 4: NETWORK
    MANAGEMENT

ETSI TBR 035
    TERRESTRIAL TRUNKED RADIO
    (TETRA); EMERGENCY ACCESS

ETSI TR 101 156

    TERRESTRIAL TRUNKED RADIO
    (TETRA); TECHNICAL REQUIREMENTS

    SPECIFICATION FOR DIGITAL
    ADVANCED WIRELESS SERVICE
    (DAWS)

ETSI TR 101 661

    TERRESTRIAL TRUNKED RADIO
    (TETRA); TECHNICAL REQUIREMENTS
    SPECIFICATION; MANAGED DIRECT
    MODE OPERATION (DMO)

ETSI TS 101 040

    TERRESTRIAL TRUNKED RADIO
    (TETRA); SECURITY; LAWFUL
    INTERCEPTION (LI) INTERFACE

ETSI TS 101 040

    TERRESTRIAL TRUNKED RADIO
    (TETRA); SECURITY; LAWFUL
    INTERCEPTION (LI) INTERFACE

ETSI TS 101 156

    TERRESTRIAL TRUNKED RADIO
    (TETRA); TECHNICAL REQUIREMENTS
    SPECIFICATION FOR DIGITAL
    ADVANCED WIRELESS SERVICE
    (DAWS)

ETSI TS 101 658

    TERRESTRIAL TRUNKED RADIO
    (TETRA); DIGITAL ADVANCED
    WIRELESS SERVICE (DAWS); LOGICAL
    LINK CONTROL (LLC) SERVICE
    DESCRIPTION

ETSI TS 101 659

> TERRESTRIAL TRUNKED RADIO
> (TETRA); DIGITAL ADVANCED
> WIRELESS SERVICE (DAWS); MEDIUM
> ACCESS CONTROL (MAC) SERVICE
> DESCRIPTION

ETSI TS 101 660

> TERRESTRIAL TRUNKED RADIO
> (TETRA); DIGITAL ADVANCED
> WIRELESS SERVICE (DAWS);
> PHYSICAL LAYER (PHY) SERVICE
> DESCRIPTION

ENHANCED DIGITAL ACCESS COMMUNICATIONS SYSTEM (EDACS)

TIA/EIA TSB
69.1-2

> ENHANCED DIGITAL ACCESS
> COMMUNICATIONS SYSTEM (EDACS) LAND
> MOBILE RADIO SYSTEM PACKET DATA
> SPECIFICATION

TIA/EIA TSB
69.3

> ENHANCED DIGITAL ACCESS
> COMMUNICATIONS SYSTEM (EDACS)
> DIGITAL AIR INTERFACE FOR: CHANNEL
> ACCESS, MODULATION, MESSAGES AND
> FORMATS

INTEGRATED DIGITAL ENHANCED NETWORK (iDEN™)

Contact Motorola Radio Systems

APCO 25

EIA IS 102.BAAA

> APCO PROJECT 25 -
> RECOMMENDED COMMON AIR
> INTERFACE - NEW

EIA IS 102.BAEC
>   APCO PROJECT 25 CIRCUIT
>   DATA SPECIFICATION NEW
>   TECHNOLOGY

EIA SP 3633
>   APCO PROJECT 25 VOCODER
>   DESCRIPTION

EIA TSB 102
>   APCO PROJECT 25 SYSTEM
>   AND STANDARDS DEFINITION

EIA TSB 102.AAAB
>   APCO PROJECT 25 - SECURITY
>   SERVICES OVERVIEW - NEW
>   TECHNOLOGY STANDARDS
>   PROJECT - DIGITAL RADIO
>   TECHNICAL STANDARDS

EIA TSB 102.AABA
>   APCO PROJECT 25 TRUNKING
>   OVERVIEW

EIA TSB 102.AACA
>   APCO PROJECT 25 -
>   OVER-THE-AIR-REKEYING
>   (OTAR) PROTOCOL - NEW
>   TECHNOLOGY STANDARDS
>   PROJECT - DIGITAL RADIO
>   TECHNICAL STANDARDS

EIA TSB 102.BAAA
>   APCO PROJECT 25 -
>   RECOMMENDED COMMON AIR
>   INTERFACE

EIA TSB 102.BAAB
>   APCO PROJECT 25 COMMON
>   AIR INTERFACE CONFORMANCE
>   TEST

EIA TSB 102.BAAC
         APCO PROJECT 25 - COMMON
         AIR INTERFACE RESERVED
         VALUES

EIA TSB 102.BAAD
         APCO PROJECT 25 COMMON
         AIR INTERFACE OPERATIONAL
         DESCRIPTION FOR
         CONVENTIONAL CHANNELS

EIA TSB 102.BAEA
         APCO PROJECT 25 DATA
         OVERVIEW NEW TECHNOLOGY
         STANDARDS PROJECT DIGITAL
         RADIO TECHNICAL STANDARDS

EIA TSB 102.BAEB
         APCO PROJECT 25 PACKET
         DATA SPECIFICATION NEW
         TECHNOLOGY STANDARDS
         PROJECT DIGITAL RADIO
         TECHNICAL STANDARDS

EIA TSB102.AABA
         APCO PROJECT 25 TRUNKING
         OVERVIEW

TIA/EIA 102 SERIES
         TELECOMMUNICATIONS, LAND
         MOBILE COMMUNICATIONS
         (APCO/PROJECT 25)

TIA/EIA TSB 102.AAAB
         APCO PROJECT 25 - SECURITY
         SERVICES OVERVIEW - NEW
         TECHNOLOGY STANDARDS
         PROJECT - DIGITAL RADIO
         TECHNICAL STANDARDS

TIA/EIA TSB 102.AABB
         APCO PROJECT 25 - TRUNKING
         CONTROL CHANNEL FORMATS

TIA/EIA TSB 102.AABC

> APCO PROJECT 25 - TRUNKING
> CONTROL CHANNEL
> MESSAGES - NEW
> TECHNOLOGY STANDARDS
> PROJECT - DIGITAL RADIO
> TECHNICAL STANDARDS

TIA/EIA TSB 102.AABF

> APCO PROJECT 25 - LINK
> CONTROL WORD FORMATS AND
> MESSAGES - NEW
> TECHNOLOGY STANDARDS
> PROJECT - DIGITAL RADIO
> TECHNICAL STANDARDS

TIA/EIA TSB
102.AABG

> APCO PROJECT 25
> CONVENTIONAL CONTROL
> MESSAGES NEW TECHNOLOGY
> STANDARDS PROJECT DIGITAL
> RADIO TECHNICAL STANDARDS

TIA/EIA TSB 102.AACA

> APCO PROJECT 25 -
> OVER-THE-AIR-REKEYING
> (OTAR) PROTOCOL - NEW
> TECHNOLOGY STANDARDS
> PROJECT - DIGITAL RADIO
> TECHNICAL STANDARDS

TIA/EIA TSB 102.BAAC

> APCO PROJECT 25 COMMON
> AIR INTERFACE RESERVED
> VALUES

TIA/EIA TSB 102.BABD

> APCO PROJECT 25 VOCODER
> SELECTION PROCESS

TIA/EIA-681

> APCO PROJECT 25 VOCODER
> DESCRIPTION

TIA/EIA/IS-102.BAAA
>   APCO PROJECT 25 -
>   RECOMMENDED COMMON AIR
>   INTERFACE - NEW
>   TECHNOLOGY STANDARDS
>   PROJECT - DIGITAL RADIO
>   TECHNICAL STANDARDS

TIA/EIA/IS-102.BABA
>   APCO PROJECT 25 VOCODER
>   DESCRIPTION

TIA/EIA/IS-102.BAEC
>   APCO PROJECT 25 CIRCUIT DATA
>   SPECIFICATION NEW TECHNOLOGY
>   STANDARDS PROJECT DIGITAL
>   RADIO TECHNICAL STANDARDS

GENERAL LAND MOBILE RADIO

AFI 33-106
>   MANAGING HIGH FREQUENCY RADIOS,
>   LAND MOBILE RADIOS AND THE
>   MILITARY-AFFILIATE RADIO SYSTEM

AFR 700-18
>   LAND MOBILE RADIO SYSTEMS
>   MANAGEMENT

CCIR V8.3
>   MOBILE SATELLITE SERVICES
>   (AERONAUTICAL, LAND, MARITIME,
>   MOBILE AND RADIO DETERMINATION) -
>   AERONAUTICAL MOBILE SERVICE

EIA SP 3215
>   MINIMUM STANDARDS FOR
>   PORTABLE/PERSONAL RADIO
>   TRANSMITTERS, RECEIVERS AND
>   TRANSMITTER/RECEIVER COMBINATION
>   LAND MOBILE COMMUNICATIONS FM OR
>   PM EQUIPMENT, 25 TO 100 MHZ

EIA-450

 STANDARD FORM FOR REPORTING
 MEASUREMENTS OF LAND MOBILE,
 BASE STATION AND
 PORTABLE/PERSONAL RADIO
 RECEIVERS IN COMPLIANCE WITH FCC
 PART 15 RULES

EIA/TIA-316

 MINIMUM STANDARDS FOR
 PORTABLE/PERSONAL RADIO
 TRANSMITTERS, RECEIVERS AND
 TRANSMITTER/RECEIVER
 COMBINATIONS LAND MOBILE
 COMMUNICATIONS FM OR PM
 EQUIPMENT 25 - 1000 MHZ

ETS 300 086

 RADIO EQUIPMENT AND SYSTEMS
 (RES); LAND MOBILE SERVICE;
 TECHNICAL CHARACTERISTICS AND
 TEST CONDITIONS FOR RADIO
 EQUIPMENT WITH AN INTERNAL OR
 EXTERNAL RF CONNECTOR INTENDED
 PRIMARILY FOR ANALOGUE SPEECH

ETS 300 113

 RADIO EQUIPMENT AND SYSTEMS
 (RES); LAND MOBILE SERVICE;
 TECHNICAL CHARACTERISTICS AND
 TEST CONDITIONS FOR RADIO
 EQUIPMENT INTENDED FOR THE
 TRANSMISSION OF DATA (AND SPEECH)
 AND HAVING AN ANTENNA CONNECTOR

ETS 300 219

 RADIO EQUIPMENT AND SYSTEMS
 (RES); LAND MOBILE SERVICE;
 TECHNICAL CHARACTERISTICS AND
 TEST CONDITIONS FOR RADIO
 EQUIPMENT TRANSMITTING SIGNALS TO
 INITIATE A SPECIFIC RESPONSE IN THE
 RECEIVER

ETS 300 230
> RADIO EQUIPMENT AND SYSTEMS
> (RES); LAND MOBILE SERVICE; BINARY
> INTERCHANGE OF INFORMATION AND
> SIGNALLING (BIIS) AT 1200 BIT/S (BIIS
> 1200)

ETS 300 230
> RADIO EQUIPMENT AND SYSTEMS
> (RES); LAND MOBILE SERVICE; BINARY
> INTERCHANGE OF INFORMATION AND
> SIGNALLING (BIIS) AT 1200 BIT/S (BIIS
> 1200)

ETS 300 279
> RADIO EQUIPMENT AND SYSTEMS
> (RES); ELECTROMAGNETIC
> COMPATIBILITY (EMC) STANDARD FOR
> PRIVATE LAND MOBILE RADIO (PMR)
> AND ANCILLARY EQUIPMENT (SPEECH
> AND/OR NON-SPEECH)

ETS 300 296
> RADIO EQUIPMENT AND SYSTEMS
> (RES); LAND MOBILE SERVICE;
> TECHNICAL CHARACTERISTICS AND
> TEST CONDITIONS FOR RADIO
> EQUIPMENT USING INTEGRAL
> ANTENNAS INTENDED PRIMARILY FOR
> ANALOGUE SPEECH

ETS 300 341
> RADIO EQUIPMENT AND SYSTEMS
> (RES); LAND MOBILE SERVICE;
> TECHNICAL CHARACTERISTICS AND
> TEST CONDITIONS FOR RADIO
> EQUIPMENT USING AN INTEGRAL
> ANTENNA TRANSMITTING SIGNALS TO
> INITIATE A SPECIFIC RESPONSE IN THE
> RECEIVER

ETS 300 390

> RADIO EQUIPMENT AND SYSTEMS
> (RES); LAND MOBILE SERVICE;
> TECHNICAL CHARACTERISTICS AND
> TEST CONDITIONS FOR RADIO
> EQUIPMENT INTENDED FOR THE
> TRANSMISSION OF DATA (AND SPEECH)
> AND USING AN INTEGRAL ANTENNA

ETS 300 471

> RADIO EQUIPMENT AND SYSTEMS
> (RES); LAND MOBILE SERVICE; ACCESS
> PROTOCOL, OCCUPATION RULES AND
> CORRESPONDING TECHNICAL
> CHARACTERISTICS OF RADIO
> EQUIPMENT FOR THE TRANSMISSION
> OF DATA ON SHARED CHANNELS

ETS 300 793

> RADIO EQUIPMENT AND SYSTEMS
> (RES); LAND MOBILE SERVICE;
> PRESENTATION OF EQUIPMENT FOR
> TYPE TESTING

ETSI EN 300 279

> ELECTROMAGNETIC COMPATIBILITY
> AND RADIO SPECTRUM MATTERS
> (ERM); ELECTROMAGNETIC
> COMPATIBILITY (EMC) STANDARD FOR
> PRIVATE LAND MOBILE RADIO (PMR)
> AND ANCILLARY EQUIPMENT (SPEECH
> AND/OR NON- SPEECH)

ETSI EN 300 279
DRAFT

> ELECTROMAGNETIC COMPATIBILITY
> AND RADIO SPECTRUM MATTERS (ERM);
> ELECTROMAGNETIC COMPATIBILITY
> (EMC) STANDARD FOR PRIVATE LAND
> MOBILE RADIO (PMR) AND ANCILLARY
> EQUIPMENT (SPEECH AND/OR NON-
> SPEECH)

ETSI EN 301 166

> ELECTROMAGNETIC COMPATIBILITY
> AND RADIO SPECTRUM MATTERS (ERM);
> LAND MOBILE SERVICES; TECHNICAL
> CHARACTERISTICS AND TEST
> CONDITIONS FOR RADIO EQUIPMENT
> FOR ANALOGUE AND/OR DIGITAL
> COMMUNICATION (SPEECH AND/OR
> DATA) AND OPERATING ON
> NARROWBAND

ETSI EN 301 166
DRAFT

> ELECTROMAGNETIC COMPATIBILITY
> AND RADIO SPECTRUM MATTERS (ERM);
> LAND MOBILE SERVICES; TECHNICAL
> CHARACTERISTICS AND TEST
> CONDITIONS FOR RADIO EQUIPMENT
> FOR ANALOGUE AND/OR DIGITAL
> COMMUNICATION (SPEECH AND/OR
> DATA) AND OPERATING ON
> NARROWBAND

ETSI TS 101 348

> DIGITAL CELLULAR
> TELECOMMUNICATIONS SYSTEM
> (PHASE 2+); GENERAL PACKET RADIO
> SERVICES (GPRS); INTERWORKING
> BETWEEN THE PIBLIC LAND MOBILE
> NETWORK (PLMN) SUPPORTING GPRS
> AND PACKET DATA NETWORKS (PDN)
> (GSN 09.61 VERSION 6.3.0 RELEASE
> 1997)

ITU-T D.93

> CHARGING AND ACCOUNTING IN THE
> INTERNATIONAL LAND MOBILE
> TELEPHONE SERVICE (PROVIDED VIA
> CELLULAR RADIO SYSTEMS)

MPT 1327

> A SIGNALLING STANDARD FOR
> TRUNKED PRIVATE LAND MOBILE RADIO
> SYSTEMS

MPT 1362

> CODE OF PRACTICE FOR INSTALLATION
> OF MOBILE RADIO EQUIPMENT IN LAND
> BASED VEHICLES

TIA/EIA TSB 69

> A SYSTEM AND STANDARDS
> DEFINITION FOR A DIGITAL LAND
> MOBILE RADIO SYSTEM

Appendix III - Associations

American Association of State Highway and Transportation Officials
444 North Capitol St., NW, Ste. 249
Washington, DC 20001
P: 202-624-5800
F: 202-624-5800
Covers highway maintenance radio service

American Automobile Association / National Emergency Road Services
1000 AAA Drive, Mailspace 15
City of Heathrow, FL 32746-5063
P: 407-444-7000
F: 407-444-7380
Covers automobile emergency radio service

American Hospital Association / American Society for Hospital Engineering / Telecommunications Section
840 North Lake Shore Dr.
Chicago, IL 60611
P: 312-280-5223
F: 312-280-6786
Covers medical radio service

American Petroleum Institute Information Systems
1220 L. St.
Washington, DC 20005
P: 202-682-8364
F: 202-682-8521
Covers petroleum radio service

American Trucking Association, Inc.
2200 Mill Rd.
Alexandria, VA 22314
P: 703-683-1934
Covers motor carrier radio service

Applied Business Telecommunications (ABT)
PO Box 5106
San Ramon, CA 94583
P: 510-606-5150
F: 510-607-9410
web: www.abctelecon.com

Associated Public Safety Communications Officers, Inc. (APCO)
1040 South Ridgewood Ave., Ste. 202
South Daytona, FL 32119
P: 800-949-2726
F: 904-322-2502
Covers public safety and special emergency radio services

Association for Local Telephone Services (ALTS)
888 17th St., Ste. 900 NW
Washington, DC 20006
P: 202-969-2587
F: 202-969-2581
web: www.alts.com

Association of American Railroads
50 F St., NW
Washington, DC 20001
P: 202-639-2217
Covers railroad radio service

Association of Public Safety Communications Officials Incorporation
2040 South Ridgewood Ave.
South Daytona, FL 32119
P: 904-322-2500
F: 904-322-2502
web: www.apcointl.com

Canadian Wireless Telecommunications Assocation
275 Slater St., Suite 500
Ottawa, ON Canada
P: 613-233-4888
F: 613-233-2032
web: www.cwta.com

Cellular Telecommunications Industry Association
1990 M St., NW, Ste. 610
Washington, DC 20036
P: 202-785-0081
F: 202-785-0721
Covers cellular radio systems

Cellular Telecommunications Industry Association (CTIA)
1250 Connecticut Ave NW
Suite 200
Washington, DC 20036
P: 202-785-0721
F: 202-785-0721
web: www.wow-com.com

Communications Marketing Association (CMA)
2824 South Kenton Ct.
Aurora, CO 80014
P: 303-576-9475
F: 303-371-8158
web: www.commktga.com/cma.htm

Energy Telecomm and Electrical Association (ENTELEC)
P.O Box 639
Tomball, TX 77377-0639
P: 281-357-8700
F: 281-357-8777
e-mail: entelec

Forestry Industries Telecommunications (FIT)
871 Country Club Road, Suite A
Eugene, OR 97401
P: 541-485-8441
F: 541-485-7556
web: www.landmobile.com

Forestry-Conservation Communications Association (FCCA)
444 N. Capitol St. NW, Suite 540
Washington, DC 20001
P: 202-624-5416
F: 202-624-5407

IEEE Vehicular Technology Society
c/o IEEE Headquarters
345 East 47th St.
New York, NY 10017

Industrial Telecommunications Association (ITA)
1110 North Glebe Rd.
Suite 500
Arlington, VA 22201-5720
P: 703-528-5115
F: 703-524-1074
web: www.ita-relay.com

International Association of Chiefs of Police (IACP)
515 North Washington St.
Alexandria, VA 22314
P: 703-836-6767
F: 703-836-4543
web: www.theiacp.org.com

International Mobile Telecommunications Association (IMTA)
1150 18th St. NW, Ste. 250
Washington DC 20036
P: 202-331-7773
F: 202-331-9062
web: www.imta.org

International Radio Consultative Committee, The (CCIR)
The CCIR is one of the ITU's permanent organs. The CCIR acts primarily to improve radio communications by establishing technical specifications. It works principally through permanently established study groups comprised mostly of volunteer experts from many countries.

International Municipal Signal Association (IMSA)
PO Box 539
Newark, NY 14513
P: 315-331-2182
F: 315-331-8205
Covers fire and special emergency radio service

International Telecommunications Union (ITU)

A treaty organization affiliated with the United Nations. Its charter includes the regulation and use of the radio spectrum. Its policies and procedures are established through a Plenipotentiary Conference, which in turn delegates much of its responsibility to a 29-member administrative council that meets annually and acts for and on behalf of the Plenipotentiary conference. World Administrative Radio Conferences for all members are called on special issues as they arise.

International Taxicab and Livery Association (ITLA)

3849 Farragut Ave.
Kensington, MD 20895
P: 301-946-5702
F: 301-946-4641
web: www.landmobile.com
Covers taxicab radio service

International Telegraph and Telephone Consultative Committee (CCITT)

The CCITT is similar to the CCIR but deals with telephone and telegraph communications.

Land Mobile Communications Council (LMCC)

c/o Keller and Heckman
West 1001 G St., NW, Ste. 500
Washington, DC 20001
P: 703-434-4100
F: 703-434-4646
An association of land mobile radio user groups and equipment manufacturers

Mfrs. Radio Frequency Advisory Committee (MRFAC)

1041 Sterling Rd., Ste. 106
Herndon, VA 20171
P: 703-318-9206
F: 703-f318-9209

National Association of Business and Education Radio (NABER)

1501 Duke St., Ste. 200
Alexandria, VA 22314
P: 703-739-0300
F: 703-836-1608
Covers business, education and specialized mobile radio services (SMRS)

National Association of State Telecommunications Directors (NASTD)
Ironworks Pike, PO Box 1190
Lexington, KY 40578
P: 606-231-1900
F: 606-231-1970
Covers statewide telecommunications

National Emergency Number Association (NENA)
47849 Papermill Rd.
Coshocton, OH 43812-9724
P: 800-322-3911
F: 614-622-2090
web: www.nena9-1-1.org
web: www.entelec.com

National Public Safety Planning Advisory Committee (NPSPAC)
This nationwide public safety plan has been implemented in the United States for the 821-824 MHz and 866-869 MHz bands.

National Telecommunications Information Administration (NTIA)
U.S. Department of Commerce
14th St. and Constitution Ave., NW
Washington, DC 20230
This is the United States Government's telecommunications policy office. It also assigns and administers frequencies to federal agencies.

Personal Communications Industry Association (PCIA)
500 Montgomery St., Ste. 700
Alexandria, VA 22314-1561
P: 703-739-0300
F: 703-836-1608
web: www.pcia.com

Site Owners & Managers Association (SOMA)
500 Montgomery St., Ste. 700
Alexandria, VA 22314-1561
P: 703-739-0300
F: 703-836-1608
web: www.pcia.com

Small Businesses in Telecommunications (SBT)
1835 K St. NW, Ste. 650
Washington, DC 20006
P: 1-202-223-8728
F: 1-202-659-0071
sso.org

Telecommunications Industry Association (TIA)
2500 Wilson Blvd., Ste.300
Arlington, VA 22201
P: 1-(703)-907-7700
F: 1-(703)-907-7727
web: www.tiaonline.org

Telecommunications Industry Association (TIA)
2001 Pennsylvania Ave., NW, Ste. 899
Washington, DC 20006
P: 202-457-4912
F: 202-457-4939
Represents manufacturers of telecommunications equipment. TIA together with the Electronic
Industries Association (EIA) prepares standards for the telecommunications industry.

Telocator Network of America
1019 19th St., NW, Ste. 1100
Washington, DC 20036
P: 202-467-4770
F: 202-467-6987
Covers radio common carrier services

The Radio Club of America
PO Box 4075
Overland Park, KS 66204-0075

UTC, The Telecommunications Association
1140 Connecticut Ave. NW, Ste. 1140
Washington DC 20036
P: 1-202-872-0031
F: 1-202-872-1331
web: www.utc.org

Utilities Telecommunications Council (UTC)
1620 I St., NW, 515
Washington, DC 20006
P: 202-872-0030
F: 202-872-1331
Covers power radio service

Index